T0277004

Subconsciousness

Yves Agid

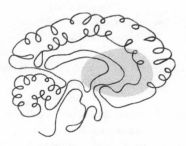

Subconsciousness

Automatic Behavior and the Brain

Columbia University Press

New York

Columbia University Press
Publishers Since 1893
New York Chichester, West Sussex
cup.columbia.edu

Library of Congress Cataloging-in-Publication Data
Names: Agid, Yves, author.
Title: Subconsciousness : automatic behavior and the brain /
Yves Agid.
Description: New York : Columbia University Press, [2021]
| Includes bibliographical references and index.
Identifiers: LCCN 2020050185 (print) | LCCN 2020050186 (ebook)
| ISBN 9780231201261 (hardback) | ISBN 9780231201278 (trade
paperback) | ISBN 9780231554015 (ebook)
Subjects: LCSH: Subconsciousness. | Consciousness.
| Human behavior—Psychological aspects. | Brain.
Classification: LCC BF315 .A36 2021 (print) | LCC BF315 (ebook)
| DDC 154.2–dc23
LC record available at https://lccn.loc.gov/2020050185
LC ebook record available at https://lccn.loc.gov/2020050186

Cover design: Lisa Hamm
Cover image: Digital composite: (top) Science Photo Library /
Alamy Stock Photo; (bottom) © Yves Agid

A Thomasine Kushner, ma très chère amie et ma voix

"And what is the use of a book," thought Alice,

"without pictures or conversation?"

—LEWIS CARROLL, *ALICE'S ADVENTURES IN WONDERLAND*

Contents

Preface

From Bench to Bedside and Vice Versa

Two streams of my life as a neurologist joined to form this book: research in the laboratory and caring for patients in the hospital. It began when, as a young resident in neurology, my ambition was to become a good neuropsychiatrist. To achieve that mission, I knew I had to go beyond what is covered in textbooks, because they present what is known, and I wanted to prepare myself to anticipate the future. I saw that we needed to stop doing medicine in what I call the "Molière way," the manner characterized by the playwright in which doctors are long on discussion but short on substance. I thought that the best way to train good neurologists was to provide the students with a high level of teaching, which entails doing excellent research that would also benefit patients—a way of thinking not in fashion at that time.

After a lengthy search, I was fortunate to find an excellent laboratory at the Collège de France in Paris, and for the next four years I devoted myself to neurobiochemistry research. When I returned to the clinic at the Salpêtrière Hospital after the years in the laboratory, I brought with me something unique at the time: double training as both a clinical neurologist and a neuroscience researcher. I never left the lab as I continued to see patients. Concurrently with clinical work, my research in my lab for over thirty-five years focused essentially on the physiological role of the basal ganglia, the structures deep within the cerebral hemisphere, the "dark basement in the brain," as described by Kinnier Wilson, that play a major role in processing the automatic motor behavior that I was observing in the examining room and, as detailed in the following pages, form part of the concept of subconsciousness.

My intention has been not to concentrate on the most understood mental cognitive functions such as memory, language, and consciousness, but instead to reinforce the neglected concept of automatic behaviors that had been studied in the past by psychologists (e.g., Pierre Janet and William James) or philosophers (e.g., Friedrich Nietzsche) under the name "subconscious," not to be confused with "unconscious" in the Freudian sense. My aim continues to be to explain what the subconscious is, its overlooked role in controlling 99 percent of our behavior, to trace its recently discovered anatomo-physiological substratum in the brain, with a special reference to one of the more ancient parts of the brain, the basal ganglia, and to examine what happens when it is ill.

The most famous disease resulting from damage to the basal ganglia is Parkinson's disease and in problems with controlling movements and posture. Patients' efforts to compensate for their loss of automatic behavior is a constant exhausting struggle. As Jean-Martin Charcot explains, "The Parkinsonian patient is condemned to produce voluntary movements for life." This prompted me to become primarily a movement disorder physician along with being a neuropsychologist.

However, research indicates that the role of the basal ganglia is not limited to movements. What we showed in my lab was that the basal ganglia also serve as the "hub" to process intellectual and emotional functions, which are all part of the concept of subconsciousness in normal persons. Further, we tried to answer the following questions: What about the concept of subconsciousness in humans compared with other animals? What distinguishes the faculty of subconsciousness from the classic concepts of consciousness and metaconsciousness? What happens when brain structures, in particular the basal ganglia, are dysfunctional, that is, when the processing of subconsciousness is abnormal?

In the examining room, I am always listening to and looking at patients very carefully. Here they are not masked by their social faces. They are totally natural and revealed as they truly are. They are giving subtle clues to their disorders, since I believe movement is the synthesis of our emotions, intellect, and physical condition.

Bench and bedside continue to merge for me: science on the one hand and observing human behavior on the other; and this extends even to ordinary life. If I look down from my window at people walking by, as part of my own subconscious, I have a feeling for who they are from their posture, gait, and body movements. Today, for example, I saw a man with a particular gait and swinging of his arm, which raised a question in my mind as to whether he might develop a serious disorder in the future. What could we learn now about the basal ganglia that could alter the course of his disease?

As we continue to understand the subconscious, it becomes clear that it is time to reconsider the basal ganglia as a therapeutic target, not just for the improvement of motor disorders but also psychiatric disorders, if, as I propose, we are to find new and more effective ways to address these diseases.

Acknowledgments

This book would not have been written without the inspiration of my wife Agnès Renard, who first suggested it. Deep appreciation goes to Timothy Daly, who provided invaluable help in translating the first version of the manuscript, and to Guillaume Palchik, whose technical skills were indispensable. For Miranda Martin, my editor at Columbia University Press, no amount of thanks is sufficient.

Subconsciousness

1

The Interrelated Levels of Consciousness

Metaconsciousness, Consciousness, and Subconsciousness and Their Relation to the Environment

It is the evening rush hour in Paris, and I am in my car driving in the Place de la Concorde (figure 1.1). A friend sits next to me in the passenger seat, and we are talking. Around us are hundreds of rushing cars, motorcycles, and bicycles. Red lights turn green and back again. Tourists are trying to cross the square in every direction. I turn my steering wheel right and left, I brake, I change gear, I accelerate. It is incredible that I have avoided an accident. Although I am being bombarded with information, it does not prevent me from being deeply engrossed in the conversation I am having with my friend. And yet, even though I am paying close attention to him, the actions required to safely navigate my car are being carried out smoothly. Indeed, without paying attention, I adapt to my environment with astonishing ease. It is as though my movements occur without my knowing or realizing—automatically, in some way. It is as if there were a hidden intelligence ensuring my adaptation to the environment.

In this sort of situation, I am not thinking about driving my car; but indeed, that is what I am doing—automatically. I have a sense of my surroundings, but I am not conscious of my actions; I am carrying them out "subconsciously."

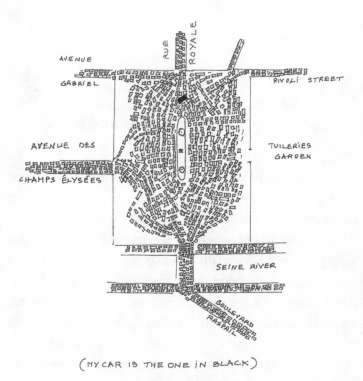

(MY CAR IS THE ONE IN BLACK)

1.1 Place de la Concorde at evening rush hour as viewed from above, with each small square representing a car.

Shortly, fighting the traffic becomes too much. I tell myself: "The traffic jam is preventing me from going anywhere. The Place is completely blocked. I've got to get out of this hell-hole." That is exactly what I do by turning off to the Champs-Elysées. As soon as I begin this move, I stop conversing with my friend. In place of an automatic activity—driving—I have just made a decision that requires intentional activity and I am no longer acting automatically. In this instance, although I am not aware that I am acting or perceiving, I am actually doing so in a nonautomatic way. I am in a state of "consciousness."

A little later, once I am out of the heavy traffic, I ask myself: "Are you crazy? You must have had a screw loose to drive around the most congested parts of Paris at six o'clock in the evening! [reflecting on my actions] That explains why I'm so on edge! [reflecting on how I feel] Yeah, you've got to be an idiot to drive in Paris at this time! [reflecting on who I am]." In my reflections, I am thinking about what I am doing, feeling, thinking and, therefore, I am in a state of "metaconsciousness." Once I am driving on the Champs-Elysées and out of the traffic jam, my subconscious takes over, and I resume the conversation with my friend.

As my frantic drive around the Place de la Concorde illustrates, the three states of consciousness are closely related: by making a decision, we pass from a subconscious state to one of consciousness in which our brain disengages from automatic behavior and involves intentional voluntary actions. When we reach the level of internal conversation with ourselves, we have moved to the state of metaconsciousness.

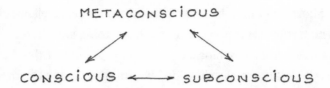

1.2 The mental functions of subconsciousness, consciousness, and metaconsciousness are interrelated.

It is important to point out that in life our state of subconsciousness is perpetual, at times consciousness dominates, and more rarely we are in a state of metaconsciousness (figure 1.2).

The Brain and the Environment

The brain is constantly receiving an enormous quantity of information, in both time (millisecond by millisecond) and space (the near and far environments). Perceptions are auditory (from background noise to close-up conversation), visual (increasingly complex due to the constant movement of both the subject relative to the environment and the environment relative to the subject), and also tactile, gustatory, and olfactory.

We might ask ourselves why our brains are not overwhelmed by being bombarded with so many signals? Fortunately, only some of this information is recorded in the brain. In a metaconscious state, even if the time spent thinking about our thoughts is reduced, there is a high degree of retained information. In a state of consciousness, there is a more protracted but reduced amount of retained information; and in a subconscious state, the retention is almost imperceptible and permanent. Subconscious automatic behavior sorts through the information constantly besieging us. It is in this way that the subconscious contributes to equanimity by taking on a role of screening autopilot.

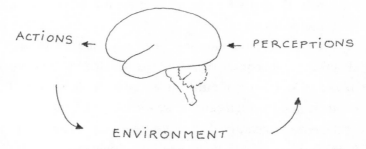

1.3 The brain's interaction with the environment.

We receive a vast amount of information from the environment, all of which is perceived by our brain (figure 1.3). These perceptions reach the posterior part of the brain and are processed to produce adapted movements. The resulting actions are modifying the environment by return.

Distinguishing the Levels of Mental Functions

A first step in distinguishing the levels of mental functions is to point out that this faculty of subconsciousness must not be confused with Sigmund Freud's notion of the unconscious.[1] For Freud, the unconscious consisted of repressed mental content that caused people's illness; and, therefore, the goal of therapy was to relieve the patient by making the unconscious conscious. Clarifying what we mean here by "subconsciousness" is difficult, because the concept tends to be neglected in medicine and philosophy. Yet it is all the more important, because recent progress in neuroscience has given new life to the way we interpret the role played by the subconscious in behavior.

When we say "the subconscious," what are we talking about? Is the subconscious a clearly defined mental function in and of itself, or is it rather an aptitude involving several components? Does it mean that we are not consciously aware of the environment, or that we do things automatically, or both? Do we constantly live in a subconscious state, or are we intermittently subconscious? Is it a uniquely human mental capacity, or is it present throughout the animal kingdom? At what moment did it appear during the evolution of life on Earth? How does the subconscious emerge during development? Can we live without it? Which brain areas are involved in the process of subconsciousness? What happens to the subconscious in the injured brain?

The following pages will attempt to provide answers to these questions by clarifying the role the subconscious plays in our lives; highlighting the physiological basis of the subconscious, with particular reference to the more ancient parts of the brain (the "basal ganglia") as compared with the more recent "cerebral cortex"; identifying brain pathologies arising from malfunctioning of the subconscious; and demonstrating the unsuspected implications for society that this—still poorly understood to this day—mental function entails.

The most instructive way to really understand the subconscious is to start by showing what it is not and how it contrasts with the other mental functions: metaconsciousness and consciousness.

Metaconsciousness: The Witness

Metaconsciousness, in the way that I shall be using this word, is a thought about a thought, an act, a perception of one's own existence. Metaconsciousness is thus the starting point of a dialogue with myself, providing the foundation for what and who I am.[2] This provides an internal reflection of my behavior toward myself, others, and the general environment. It allows me to judge and control my own acts based on a past or present situation. Metaconsciousness allows me to observe myself: I think that I think, act, and feel—that I exist. I become the object, the witness, of my own thinking. It is illustrated by the question I asked myself in the traffic jam: "Are you crazy?" How far back must we trace evolutionary history to find this "thinking witness"? It is difficult to say, for only beings endowed with language—humans—can make and share claims about metaconsciousness. What can we say about nonlinguistic animals when they recognize themselves in a mirror? The mirror test illustrates this (box 1.1).[3]

BOX 1.1: THE MIRROR TEST

When a chimpanzee, with a spot painted on its forehead, is placed in front of a mirror, it is at first intrigued by the image it sees. It inspects the mirror and then tries to see if another animal is hidden behind it. It comes back to the mirror, gesticulates, and observes these same gestures reflected. Finally, it touches the spot on its forehead: the animal has recognized itself. Except for great apes, dolphins, elephants, and corvid birds, most animals are incapable of recognizing themselves in a mirror.

Can we call recognizing that it is reflected in the mirror true meta-consciousness in the sense that the animal not only becomes aware that it is reflected in the mirror, but also that it is aware of its own existence? Is it a conscious behavior, such that, although the animal understands that it is itself in the mirror, it does not, all the same, become aware of its own existence as such?

Moreover, among humans, metaconsciousness can fade away, as is the case in Alzheimer's disease (the most common cause of the degradation of intellectual, emotional, and behavioral capacities known as dementia). It all starts with a "shrinking of the field of consciousness," that is, a diminished awareness of abstract and distant events. Then the boundary of "conscious" exploration itself narrows: the local neighborhood, the building, the apartment, the bedroom, the armchair, the bed—with all the ensuing psychological and social consequences.

What would happen if we did not have metaconsciousness? Faced with the necessity of adapting to our environment, could we survive without metaconsciousness? And if we could, would it not be forced servitude? What would be the implication for the philosophical question of personhood? Yet in daily life, how much time do we spend in this state of metaconsciousness? What would you reply if you were asked how much time on average you spend thinking that you are thinking, or that you are doing, or that you are feeling, or that you *are*? A few minutes a day? In any case, not several hours—unless you are a mystic. However, it is indeed this metaconsciousness that makes for the uniqueness and power of humans, by allowing us to control our reasoning and ourselves, which includes adjusting our behavior toward others. It is the first step in the voluntary control of mental capacities as complex as anticipation, humor, sense of fair play, simulation, and abstraction. In short, without metaconsciousness and self-reflection, there would be no ethics.

Metaconsciousness allows us to "perceive ourselves as a subject that thinks, feels, and acts."[4]

Consciousness: The Intelligent Operator

Everyone knows, in an indescribable way, the meaning of the word "consciousness"—as long as, William James so aptly said, "Nobody asks us to define it."[5] To clarify "consciousness" as it is used here, we must first eliminate ways in which the word is used in different contexts, such as "he lost consciousness" (he fainted); he acts with true professional conscience (he acts honestly and meticulously); "he has a guilty conscience" (he feels remorseful). As illustrated by the story of driving my car in the Place de la Concorde, consciousness allows us to interact voluntarily, in a nonautomatic manner, with the environment, with the help of our senses, which bring information to the brain. Thus understood, consciousness is not to be identified with knowledge (as one might be tempted to do so etymologically: cum [with] plus scio [I know]). It must be noticed that consciousness does not allow a direct access to reality but provides a reconstruction of the external world, as if it were collecting a series of clues eliciting an interpretation of that reality. Our subjective perception of the outside world, that is, our consciousness, is delayed compared with reality (for instance, to become conscious of a visual stimulus takes about 300 milliseconds, although physical perception takes between 100 and 200 milliseconds).[6]

Showered by perceptions (e.g., visual, auditory), affective states (e.g., fear, joy), and intellectual information (the choices we make every moment), the nervous system is sometimes led to make rapid, even instantaneous decisions, such as in dangerous situations. But most often, we have leisure with respect to decision making, in both relatively easy (choosing which clothes to wear) and difficult (choosing a political stance) contexts. In each case, consciousness has come into play, allowing us to determine a plan of action. "To act or not to act?" That is the question that consciousness constantly strives to answer.

Consciousness is therefore the "operator" that determines and controls our behaviors, whether simple (such as catching an object) or complex (solving a problem). But it is an intelligent operator that manages all human activities requiring adaptation to new or unusual situations. Disengaged from repetitive and automatic routine activities, this operator allows us to elaborate individual and collective strategies with others. It allows us to be open to imagination, innovation, creativity. Thus, consciousness plays a critical role for Homo sapiens, first during the process of learning in children and then in adults, for whom the time spent in a state of consciousness varies among individuals: minimally for those who spend their time performing routine tasks; substantially in the adventurer, the inventor, the thinker, the artist. In short, behavior that is not automatic is conscious, but to a varying degree among animals: absent in earthworms (figure 1.4), highly reduced in amphibians and reptiles, and at its peak in mammals.

1.4 Animal consciousness.

When during the brain's evolutionary history did consciousness appear? Logically, as soon as behaviors became no longer automatic or routine. Therefore, consciousness is operative among beings who have developed more than a simple spinal cord, that is, the first signs of a brain. This obviously includes all mammals. For the others, the aptitude of consciousness increases from the insects and amphibians (if it exists at all) through to reptiles and birds (in which we begin to see surprising cognitive capacities).

And in humans? What would happen if consciousness were reduced to a secondary role, or even disappeared? The human species would be in a situation of permanent automatism. It would paint a terrifying sociological picture of servitude. Consciousness constitutes the essence of "humanness" and allows us to survive, given our physical vulnerability compared with other animals. Thanks to consciousness, we have managed to dominate the physical world: ingenious, we make tools and weapons; cunning, we set traps; social, we make alliances; intelligent, we transmit abstract thoughts in intelligible form. In short, consciousness assures our freedom and safeguards us. In sum, consciousness behaviors are those in which we do things intentionally and nonautomatically. This suggests that our consciousness only provides a restricted general survey of the vast extent of information processed by our brains.

Subconsciousness: The Autopilot

Why, without thinking, do I put one foot in front of the other as I walk? Why can I ride a bicycle without paying attention? Why does the Place de la Concorde not become a nightmare of collisions during the evening commute? Why do people not constantly bump into one another in crowded subway stations? Is it because there are imperceptible signs indicating where others are headed? These signals are indeed present and so discreet that we hardly pay attention to them. They acquire a significance for the other person, who associates gaze, shoulder and head movement, and rapid changes of direction with meaning. They are immediately, implicitly understood by everyone, because we have become accustomed to such signals from the time we began to walk. And no one thinks of them, because they are voluntary activities, carried out without our even realizing, without being conscious. Essentially, this state is like an autopilot (figure 1.5).

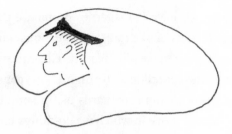

1.5 The brain on autopilot.

In fact, just as a flight captain monitors and assures the dependable functioning of the airplane in autopilot mode, the subconscious can assure the proper functioning of the brain by putting itself on "autopilot." I am using the term "auto," because an automaton is that which "moves itself, unaware, producing an activity whose only goal is to ensure the continuity of its normal functioning."[7] The word "pilot" indicates that this automaton conducts behavior. Compared to the definition of consciousness or metaconsciousness, subconsciousness (sometimes termed the "cognitive unconscious")[8] could be described as: "I do not think that I am doing, feeling, or thinking, but I am doing, feeling, thinking—in an automatic way." In that sense, the word "subconsciousness" implies the notion of both the absence of awareness and the presence of automaticity.[9] It is a kind of savoir faire, a knowing how, that has the capacity to use the information retained in our brain and to automatically adapt to innumerable events in the environment. Different from consciousness, which expresses itself in words, this "knowing how" is silent, a sort of "silent consciousness."

This subconscious autopilot is, therefore, responsible for our habits and routines. The advantage is that it facilitates daily existence by allowing consciousness to avoid boring activities: subconsciousness allows us to drive a car without difficulty, to train using repetitive movements in sport, to tend a garden, or to dance with abandon. In this way, it takes charge of necessary habits, allowing our intellectual and affective capacities to be expended in more captivating tasks. However, care must be taken, as we can become slaves of these automatic activities, which can soon become rituals. With time, the zealous person can become intolerably scrupulous and the active individual compulsively agitated. Without realizing it, automatic behavior tends to become excessively repetitive. Who among us, lost in reflection, has not automatically taken the known route to a destination, when another route would have been better? Even worse, these habits can become a nuisance, such that we cannot break free. The subconscious can take over to the point of altering our behavior. For instance, when receiving (or being subjected to) thousands of oral repetitions, the messages become mechanically, subconsciously, ours, sometimes leading to harmful and negative prejudice such as racism or untruths such as denying climate change and the effectiveness of vaccinations. However, we cannot live without the subconscious, except perhaps when we sleep.

The three mental functions—metaconsciousness, consciousness, and subconsciousness—are clearly seen in child development and become permanent features of human life. During the first months of life, newborns shake their arms and legs without an apparent goal. They seem to recognize their mothers' faces; they express joy and anger. We can probably qualify these behaviors as subconscious. From the age of three to four months, infants start to concentrate their attention on a subject of interest. Toward the age of six months, they can recognize emotions in people. At twelve months, babies adapt to people and the surrounding environment. Then they start to recognize themselves in a mirror, as though they had an elementary consciousness of themselves. But this consciousness is limited to the present; it is not yet metaconsciousness in the sense of "I am conscious of my own existence." It is, more simply, a taking account of oneself, in the sense of "I think that it's me that I see in the mirror." Toward the age of five, children really begin to become aware of their thoughts, those that concern themselves as well as other-directed thoughts. They start to lay down memories. They use "I," knowing that it refers to them. Although it was a while ago that they started to laugh, they can become aware of real humor. It is only later that children begin to distinguish truth from falsity (which can bring them to lie) and to stop being self-centered. They guess the thoughts of others and learn to conceal their own.

2

The Complexities of Subconsciousness

My unconscious knows more about the consciousness of the
psychologist than his consciousness knows about my unconscious.

—KARL KRAUS

Subconsciousness—automatic thinking, acting, and feeling without being aware—is highly complex, whether
this mental activity is intentional (with a desired goal)
or unintentional (involuntary) as shown in figure 2.1. Intentional acts are stored in memory, whereas unintentional acts
are not necessarily retained.[1]

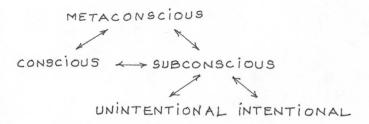

2.1 Intentional and unintentional acts are also related to the mental
functions of subconsciousness, consciousness, and metaconsciousness.

Intentional Subconsciousness Expresses a Will with a Purpose

Even when we are not thinking about what we are thinking, doing, or feeling, we are constantly undertaking highly adapted actions. They can be simple: tying my shoelaces, peeling vegetables, shaking someone's hand (e.g., the particularity of a handshake being that the other person irresistibly offers his or her hand in response as in figure 2.2), clenching my jaw while exerting effort. Or these actions can be more complex: driving my car, washing myself, taking the elevator, juggling balls, moving my lips while I speak or my fingers while I write with my fountain pen (whereas the content of what I say or write is anything but automatic), the powerful service motion of a tennis champion and the dazzling figures of a floor gymnast. These acts are indeed

2.2 A simple handshake.

automatic, thus falling under the category of subconsciousness, and they are intentional and goal oriented.

The expression of subconscious behavior is always neuromotor based, but this behavior reflects intellectual and affective activities. Intellectually, it is illustrated by the know-how of the manual worker working in a factory line, the housekeeper doing the same housework over and over again, the marching soldier, the pupil reciting the times tables, the conventional speech given by the mayor during an agricultural show, and the sales-talk of a domestic salesman. Intentional subconsciousness is therefore a kind of refuge. Being truly automatic, it carries out tasks without effort, without fear of failure, and without creative intentions. We do not have to implement it; subconsciousness imposes itself. We can think of it as a kind of preformed "know-how" dictated by our need to adapt to our environment.

We might ask to what extent these movement patterns performed time and time again contribute to the fixation of belief and the dogmatisms cluttering the human mind? They are activities performed without needing to reflect on the task at hand. At the affective level, all kinds of behavior are evident, from the irresistible smile of the mother contemplating her baby, to the growing anger provoked by a trifle, to the emotion someone feels in a familiar place. These behaviors are often both intellectual and affective; for example, if I sing (I speak words having meaning, but their expression is fashioned by the emotion I put into the song) or if I dance (a set of movements I have learned, made even more emphatic when I "put my heart" into the dance).

2.3 Love.

What if love—when we fall head over heels in love—were subconscious? Everyone has fallen in love, at least once. At such a time, you did not calculate your reasons, you did not think: "Well, he has nice eyes and his ears don't stick out." Or "she has an engaging smile even if it is a little crooked." Instead, you were "bitten by the love bug," you "fell in love." Curious, isn't it? We do not grow in love—we fall. Contained within this verb is the notion of abruptness, but also the idea of a fall that happens without your decision, as though it were independent of your will (figure 2.3). It is as though the act was automatic, without you realizing it, without you being immediately conscious or consenting. Here, we have a strong and subtle emotion—love—that seems to fall within the frame of "subconsciousness," felt by every man and woman, as well as nonhuman animals, like the pigeons, who "s'aimaient d'amour tendre" (loved each other tenderly) in the poem "Two Pigeons" by the seventeenth-century poet

Jean de la Fontaine.[2] Indeed, everybody has observed pigeons kissing each other. After several hours or days, the female finally chooses one of the numerous males that are courting her. It's like humans—except for the size and the feathers! However, as far as the brain is concerned, there are several differences between pigeons and humans: their brain is 500 times smaller. Then, the question arises: What are the physiological bases of love in the brain? Trust Brits to look for the neuronal substrate of romantic love (box 2.1).

BOX 2.1: IS ROMANTIC LOVE SUBCONSCIOUS?

Volunteers were recruited who considered themselves "truly, madly, and deeply in love."[1] Their brains were studied with magnetic functional resonance imaging. One of the members of the passionate couple (for example, the male) contemplated portraits of members of the opposite sex familiar to the couple and at intervals a picture of the loved one appeared (his fiancée, in this case). "Love" was quantified by measuring the brain metabolic metabolism provoked by the beloved compared to the known members of the opposite sex (representing the background noise of the study). Surprisingly, instead of the evolutionarily recent "emotional" areas of the cerebral cortex being activated, as one might expect for such a sophisticated emotion, the most metabolic activation was found in the basal ganglia (located in a small area in the center of the brain).

This observation shows that an emotion as subtle as "romantic" love primarily activates one of the oldest structures in the brain that is most maximally developed in reptiles and birds.

---

1. Andreas Bartels and Semir Zeki, "The Neural Basis of Romantic Love," *Neuroreport* 2000:13829–34, doi: 10.1097/00001756-200011270-00046.

The Intentional Subconscious Is Permanent

How much time do we spend each day reflecting on what we are doing and feeling, pondering who and what we are? In short, how much time do we spend in metaconsciousness? Actually, humans spend little time thinking about their thoughts, perhaps a few seconds or minutes per day—unless we consider the hermit, mystic, narcissist, or introvert. How much time do we spend doing and feeling, in such a way that we adapt correctly to the changing environment, that is, in a state of "consciousness"? Longer, certainly, several dozens of minutes, perhaps an hour or two for the more dynamic among us. But what is striking is that we are permanently in a state of subconsciousness, whether the subconscious is in play or merely performing a supporting role: adopting a posture (hunched over a computer), performing gestures (riding the clutch, accelerating, writing, playing musical scales, speaking), carrying out intellectual tasks (addition, multiplication, filling out an Excel spreadsheet), or perceiving routine noise (crowd movement in the street, the screeching subway, background music). Is it not surprising to realize that we are constantly in a state of subconsciousness during our entire lives?

On its own, subconsciousness has the capacity to direct our automatic adaptation to the environment, provided we are not facing a novel situation. Often, however, subconsciousness does not act alone. It serves an accompanying role when we engage in tasks that involve reflecting on our

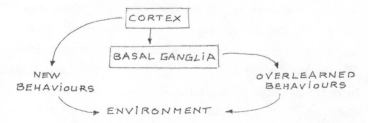

2.4 The roles of the basal ganglia in overlearned behaviors and the cerebral cortex in new behaviors. The cerebral cortex (located at the periphery of the brain), initiates new nonautomatic behavioral patterns to adapt itself to a change in the environment (consciousness), thus modifying the environment (*left*). Automatic overlearned (skills acquired beyond the point of initial mastery) motor programs (subconsciousness) are controlled by the basal ganglia (*right*). However, the basal ganglia remain permanently under the control of the cerebral cortex, so that in the case of a new situation, the cortex can disengage itself from the basal ganglia to adapt the individual to a new situation, whereas the basal ganglia continue to ensure the automatic adaptation of the individual to the environment.

own thoughts (metaconsciousness) or adapting to a novel situation (consciousness). We can distinguish the least automatic behaviors from those that are more automatic, recognizing there a constant interchange between the two, as though they lay on a continuum. When easy maintenance is the order of the day, intentional subconscious behavior takes control by itself, all the while with the constant accompaniment of the conscious, the latter intervening to change behavior as, and when, necessary; for instance, if I am smoothly walking—a subconscious behavior— I may decide to accelerate my gait—a conscious behavior.

In most situations, consciousness leaves subconsciousness completely free. When response to a novel or more useful task is needed, the automatic function of subconsciousness allows consciousness to disengage from these lifeless tasks in order to adapt its aptitudes for intelligence to a novel question (figure 2.4). This may explain why some of us, in particular philosophers such as Socrates or Epictetus, have been known to be in the habit of walking to facilitate the emergence of new ideas. There is no consciousness or meta-consciousness without subconsciousness, because the latter is always present with or without such illustrious companions. Intentional subconsciousness is thus like the foundation upon which other functions are built. It is always there, permanent and exclusive, for most, if not all, animals with a nervous system. These differences in mental functions and behaviors can be traced in brain activity.

The Intentional Subconsciousness Is Ubiquitous

If we accept that subconscious activities are observed in living beings that have a brain, then this function should be detectable in most animals. However, it will obviously be reduced to its simplest expression in the most primitive animals whose most recent phylogenetic structures are not yet present. It becomes preeminent in animals such as amphibians, fishes, reptiles, and birds. This probably explains the extraordinary motor skills of animals like frogs that can catch an insect in flight (figure 2.5), birds of prey that dive on mice, and gibbons that jump effortlessly from branch to branch. In any case, these skills are sufficient to ensure the survival of species, because animals having essentially, if not exclusively, only a subconscious appeared in evolution long before humans and are doing very well. This demonstrates the power and robustness of this ability, which has allowed the animal kingdom to survive to this day. Animals use these faculties continuously for elementary acts of existence such as hunting or gathering food, but also for the more complex acts of social life. As humans, we do the same but not as well, with the additional quality (which is in the domain of consciousness) of questioning to learn, to evaluate past and present experiences, to anticipate the future.

2.5 Fly-catching . . . It's subconscious!

The Intentional Subconscious Is Learned

For the vast majority of animals to have survived for millions of years, it was necessary to have a gene pool that was wonderfully selected by evolution. This is apparently sufficient for insects, which are probably exclusively subconscious. This is certainly not the case for more recently evolved and more highly complex animals, such as mammals, which could not survive if they had not acquired other means of defense.

For those animals, a permanent learning process, such as one needed to acquire a difficult new behavior, gradually becomes automatic. In humans, everything begins at birth and takes on full meaning as we acquire a certain independence from our parents. Our learning abilities are enriched until adolescence, in parallel with brain development. They stabilize in adulthood and slowly but surely become reduced in very old age. Once acquired, behaviors are so well learned that they become strongly fixed, which naturally facilitates their automaticity (figure 2.6). Examples include: the speed of the swift bird, which can exceed 200 kilometers per hour and spends most of its life never touching the ground, the legendary acrobatic skill of marmosets, and humans developing writing skills. This is how, once having learned to swim, you will know how to swim all your life. Similarly, a skier who has not practiced the sport for years will soon get back into the routines of the downstream or upstream turn, whereas, learning this sport at an advanced age becomes difficult. How are these behaviors learned?

2.6 Intentional subconscious activities: driving a car, walking, playing
tennis, dancing.

These intentional automatic behaviors are learned by
imitation; specifically, by "passive" imitation so that, unin-
tentionally, almost without our knowledge, our behavior is
permeated by the person being imitated. This way of learning, by copying without realizing it, plays a primordial role
in our formation as individuals. At the beginning, the baby
observes and adopts the gestures of his or her mother;
later, to a lesser degree, he or she acquires the accent of a
foreign language, perhaps soaks up the refined behavior of
an admired relative, learns the techniques of the worker, or
molds himself or herself according to the model of a vener-
ated teacher. This is how an individual's personality is shaped
by the environment. This passive imitation is different from
the "active" imitation of schoolchildren, whose proficiency
is enriched by the teacher, and from the actor who repeats
a text with appropriate facial expressions. This contribution
to education is more in the domain of the conscious mind.
Thus, when we are small, we are taught to behave well, pos-
sibly to adopt cultural behaviors, to acquire manual skills,

even ingenuity to read, write, and calculate. Although these teachings are based on verbal interactions, perhaps even more impactful in shaping us is the continuous perception of the behavior of those we imitate.

Repetition is necessary to achieve the kind of dexterity that no longer needs reflection; but the task becomes more difficult when learning a new motion requires reflection. And the task becomes even harder when it is necessary to correct and replace already acquired patterns. Intense intellectual concentration is needed to adapt the new movements so that they become routine. However, beyond even the most in-depth learning, the quality of motor performance depends on a subtle finesse that the art of teaching cannot provide. This is where the intrinsic qualities of the craftsperson who becomes an artist, the athlete who becomes a champion, and the researcher who makes a discovery play a crucial role. In short, something extra intervenes to make the ace or the genius.

Once acquired, these elements learned with effort (for example, playing a musical instrument, singing, or dancing) form a foundation for reproducible behavior. They truly form a subconscious memory. If the behaviors are learned well, they can sometimes successfully mask deficiencies. This is seen in the case of educated people in old age when their intellectual faculties are diminished or even more explicitly in patients with Alzheimer's disease who maintain basic social behaviors ("How are you?" . . . "I am good." . . .

"Please give me the salt." . . . "Take care."), although they have already lost the core of their memory.

Intentional Subconscious Is a Reflection of the Personality

If subconsciousness is a way of thinking, doing, feeling, and being without our knowledge, then considering the state of subconsciousness in others should provide us with information about what they think, feel, and really are. Because people are in a permanent state of subconsciousness, careful observations are revealing. We are expressing our unguarded thoughts and emotions in real time—that is to say, the intimacy of what is deepest inside ourselves—for better or worse. Our innermost selves can be expressed in the charm of someone who naturally knows how to create sympathy by putting others at ease; or by another who, after strutting around in public, reveals his or her true nature in a moment of relaxation. This observation of subconscious behaviors is particularly instructive because, being natural and irrepressible, they cannot be hidden. They are easy to observe, because they are always present in one way or another.

The way a patient walks offers a window through which psychologists and physicians can view that patient's personality. Bodily movements and facial expression are also what make the fortune-teller so successful. The seer (sometimes in good faith, but not always) uses their talent to guess the customer's background. To be convincing, and thus impressive, the fortune-teller's powers of observation must be strong enough to break down the myriad components of the client's behavior.

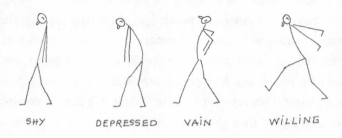

SHY   DEPRESSED   VAIN   WILLING

2.7 Walking reveals emotional states.

Observing the general appearance of a person—his or her way of walking (figure 2.7), body language, and facial expression—is enough to tell us who we are dealing with. It is above all the expression on the face that makes it possible to immediately recognize the tormented or joyful, the distracted or scrupulous, the timid or the extroverted. The reason is that the contraction of some of the dozens of muscles that animate the face results in microexpressions such that everything is revealed—even lies. The power of the automatic motor program on our face is such that the nonautomatic expression (consciousness) of an idea that we want to hide (by pretending) or of an emotion that we want to conceal (for example, stage fright) cannot always be masked. In fact, it is very difficult to pretend (except for some serious criminals who end up convincing themselves that they are innocent, as encountered by the fictional detective Maigret in the novels of Belgian author Georges Simenon), because "le naturel revient au galop" (what is natural always returns swiftly). With extreme effort of consciousness, we can pretend; but it is difficult to bluff our subconscious, which is a naturalness from which we cannot escape. Look and listen to the person and his or her subconscious will show you who that person really is. The best reflection of what is happening in the brain is observed (facial expressions and body posture) and heard (timbre, tone, intensity).

The Sick Subconscious: Abnormal Involuntary Movements, Thoughts, and Emotions

Abnormal Involuntary Movements

This man trembles when he pours water into his glass, although, strangely enough, at rest he does not tremble (he has "essential tremor"). Another person does not tremble in his physical bearing, but his fingers exhibit a slow tremor at rest (he has Parkinson's disease) (figure 2.8).

Still another person has small, brief, explosive, disseminated movements of the extremity of the limbs (chorea observed in patients with Huntington's disease). One young man is constantly agitated by twisting movements of the four

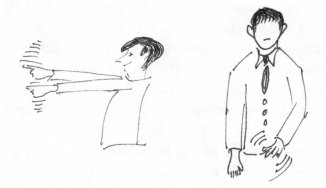

2.8 Postural (*left*) and rest (*right*) tremor. The postural tremor (*left*) of rapid rhythm (10–12 cycles per second), observed in the postural attitude (disappearing mostly at rest), has the characteristics of the essential tremor (frequent, rarely very disabling, relatively benign). The resting tremor (*right*), of slow rhythm (three to six cycles per second), at the end of the upper limb (most often unilateral at first), is characteristic of Parkinson's disease (which is accompanied by other symptoms, including slowness and "cogwheel" rigidity).

limbs and the trunk, causing strange postures (dystonia) (figure 2.9). He clears his throat and makes sudden rightward movements from his neck but says he can prevent his bursts of abnormal movements by an act of will (he has tics). These are all excessive movements, true involuntary abnormal movements that interfere with normal motor skills. Abnormal movements that indeed occur subconsciously, as they appear irrepressibly, independently of his will, and if he becomes aware of it, it is because these dyskinesias are embarrassing or even disabling.

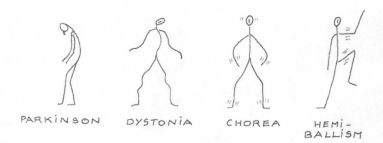

2.9 Other abnormal involuntary movements that are part
of the diseased subconsciousness.

Abnormal Involuntary Thoughts and Emotions

The situation is not the same for an intellectual disorder, constituting what could be called "involuntary abnormal thoughts" (thoughts that are always ultimately expressed by movement). These are poorly formed, poorly controlled, inappropriate thoughts, with the specific quality that they appear without the patient's knowledge. They are characterized by intellectual slowness, apathy, or conversely, intellectual excitement, a flight of ideas, intrusive thinking, what we refer to in French as "un petit vélo dans la tête" (the little bicycle in the head) as in figure 2.10.

The same qualification can apply to emotional disorders, true "involuntary abnormal emotions." These can range from anxiety and depression, to obsessive-compulsive disorders, and even delusions. These psychiatric symptoms meet the definition of subconsciousness, because they occur

2.10 The little bicycle in the head.

without the patient's knowledge, realization, or volition, in an unintentional and therefore automatic way. Basically, they respond to the definition of involuntary abnormal thoughts and involuntary abnormal emotions, as it is the case with classic involuntary abnormal movements.

These symptoms (emotional, intellectual, or motor) may occur in isolation or in combination, in which case they produce a global behavioral disorder. The fact that they appear automatically raises the question of the role of brain structures involved in controlling these symptoms. It makes sense that they should result from the dysfunction of neural networks involved in controlling movement, intellect, and emotions in the brain. However, the structures involved in these disorders originate in segregated emotional (referred to as "limbic"), intellectual (referred to as "associative"), and motor circuits. When these symptoms are automatic-involuntary (e.g., depression in Parkinson's disease, intellectual disorders in Huntington's disease), the dysfunction occurs mainly in the basal ganglia. When the lesions are restricted to the cerebral cortex, as in frontal dementia, automatic abnormal behaviors appear as if the basal ganglia, freed from the instructions given to them by the cortex, were no longer controlled. More detail regarding the major role and importance of the basal ganglia are explored below.

Unintentional Subconsciousness: With and Without Memory

As mentioned earlier, the amount of time spent on knowledge of one's own thoughts at a given moment (which defines the state of consciousness) is very small in most individuals. The same is true, to a lesser degree, of the state of consciousness that characterizes the period during which an individual engages in an intellectual activity (to solve a problem, for example) or focuses on sensory perceptions (such as becoming familiar with a new melody). This means that most of life takes place in a subconscious state. When this type of consciousness expresses a will with a purpose, it is intentional. In some cases, however, there is no will and no purpose. It can be said to be unintentional: "I think, I act, I feel, automatically, without thinking about it, but this time without intention." The question regarding the unintentional subconscious is whether or not I remember the content of this unintentional subconsciousness.

Unintentional Subconsciousness with Memory

I am taking a walk, not thinking about anything in particular. Suddenly, around the corner, I come face to face with a lion (figure 2.11). I am petrified, and without having time to ask myself questions, I run away. I am not thinking, "Oh, a feline. He might eat me. What to do about it? It's better to run away!" I am running away before I even think about doing it. Instantly, I run. It is automatic. Automatic but not intentional; and fortunately so, because if I had to stop and ask myself that question, I would not be here to tell the tale. But do I remember the encounter? You bet I do!

2.11 Face to face with the lion.

The lion example illustrates an almost instantaneous automatic behavior: no time to reflect, but a precise memory. Given the intensity of the reaction, this behavior could be considered a so-called reflex action. But it is not a reflex, which results in a rapid response from a specific place of stimulation through the spinal cord (e.g., the patellar knee-jerk reflex). Such an elementary reflex alone cannot cause the complex behavior of flight. The functions of metaconsciousness and consciousness did not have time to intervene in this mental procedure. Because violent emotions are at the root of automatic escape behaviors, the basal ganglia, brain structures at the center of the brain that play a prominent role in automatic behaviors, are plausible candidates to explain this phenomenon of unintentional subconsciousness. It remains to be demonstrated, but it is a difficult experiment to carry out—not all laboratories have a lion.

Unintentional Subconsciousness Without Memory

Then there are times when, "I thought, I did, I felt, but without paying any particular attention or realizing it—in short, automatically—I don't remember it." And how can I talk about it if I do not remember? The difficulty lies in the fact that I must show that I have stored information even though I have no memory of it. The information may not have been sufficiently acquired due to lack of attention (this is the case with distraction); it may also not have been recorded (for example, in a patient with Alzheimer's disease). Apart from these particular cases, experience makes it possible to distinguish two distinct situations, depending on whether or not recall is possible.

Situation 1: Infraconsciousness

There is information that has certainly been encoded, but the subject is not aware of it. This is the case, for example, of patients with a massive visual perception disorder that severely affects a restricted territory of the posterior cerebral cortex. The patient no longer sees anything, but it can be shown that he or she has a certain form of visual perception but is unaware of it (figure 2.12). This is the classic phenomenon of "blind vision."[3] This blind person is "blinder than a blind man," for he or she not only does not see, but refuses to admit to being blind (anosognosia); the patient is not aware of

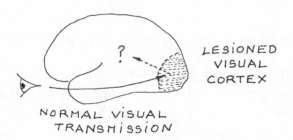

2.12 The blind man "more blind than the blind man." Bilateral lesions of the occipital lobe (hatched area), which integrates visual perceptions, usually resulting from an infarct by obliteration of the two posterior cerebral arteries, leads to "cortical blindness." Although the visual pathways from the retinas to the cerebral cortex are intact, the patient has lost his sight, most often without really being aware of it. However, all other mental activities, automatic or not, are preserved.

being blind. And yet, it can be shown by using sophisticated experimental devices that traces of what has been perceived have been encoded in the brain. It is indeed subconscious.

Such subconscious phenomena are also observed in normal subjects; for example, when a word or image is presented for a period of time so brief that the subject is certain not to have seen it,[4] as in figure 2.13. Such subliminal techniques

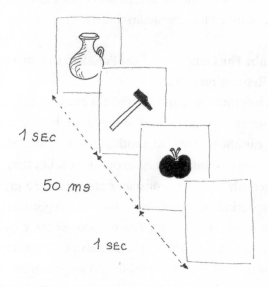

2.13 Subliminal vision. A control subject is placed in a magnetic resonance imaging machine to detect brain areas where metabolic activation occurs. Images (first a carafe, then an apple) are presented to the subject about every second for the subject to remember. When an image is presented for a sufficiently shorter time, less than 200 milliseconds (here a hammer), the subject does not remember it, having not had enough time to become aware of it. However, with the help of special devices, it is possible to show that the subject has kept a memory trace of this ultrashort presentation.

have been used to make propaganda (images presented for less than 50 milliseconds are not consciously perceived, but they leave traces such that subjects discern without realizing it). What is true, roughly speaking for propaganda, is also true for advertising; this has led to the banning of subliminal advertising on television. In this form of subliminal subconsciousness, called here "infraconsciousness," it can be shown scientifically that the information has left a trace in the brain, although the information does not reach consciousness, much less metaconsciousness.

Situation 2: The Dungeons of the Freudian Unconscious— Psychic Repression

There is information acquired without our being aware of it; it has been stored somewhere in the brain, has been forgotten and cannot be recalled spontaneously. These thoughts and emotions are buried deep in our brains, but they cannot spontaneously return to consciousness despite a great deal of introspection. It is nothing more than repression in the Freudian sense of the term. The purpose of the psychoanalytical approach is to bring to consciousness the tormenting thoughts that have been set aside and forgotten but remain present somewhere in the brain. These are "repressed" thoughts, buried in the "dungeons" of the brain. The person has apparently forgotten previous painful events; he or she does not remember them. This is the case of unintentional subconsciousness without memory. The objective of the psychoanalytical treatment is to bring the patient's hidden thoughts out of oblivion and subtly assist in drawing them into the field of consciousness so that they can be taken into account. If this description is correct, it could be suggested that in therapeutic analysis sessions, the patient would move from the stage of unintentional subconsciousness without memory, to the stage of intentional subconsciousness, and from there to consciousness (figure 2.14). Conversely, the genesis of repression would have occurred by passing from the stage of consciousness to that of the "Freudian unconscious," the psychoanalysis session allowing the patient, in principle, to change to moving in the opposite direction.

This is also the case with dreams, with their chaotic scenes that can reflect certain painful emotional states, and the "Freudian slip," such as the word with a hidden meaning or released inappropriately (during a lapse).

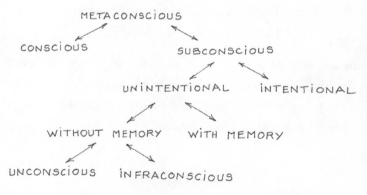

2.14 The Freudian unconscious. This figure illustrates a model of mental functioning that indicates that, beyond psychoanalytical interpretations, metaconsciousness, consciousness, subconsciousness, and Freudian unconsciousness are linked in top-down and bottom-up information processing.

In short, memory is either recallable, as in the Freudian unconscious, so that repressed thoughts can reemerge and the unintentional subconscious becomes intentional again (then conscious or even metaconscious); or unrecallable, even if it has left a trace in the brain, it is the infraconsciousness of the kind tapped into by unauthorized propaganda or advertising. In any case, we are well within the framework of subconsciousness (we do not think *about* it, it is automatic), unintentional (because there is no deliberate will to keep the thing in memory) and without memory (although in the case of psychoanalytic repression, it is possible to bring up forgotten scenes). There is therefore a constant flow back and forth between the different forms of subconsciousness and consciousness.

From Unintentional Subconsciousness to Consciousness and Vice Versa

In addition to the Freudian unconscious, there are other situations in which we move from a subconscious state to a conscious state, as suggested by three examples: benign forgetting, hysteria, and intuition.

Benign Forgetting

I meet an acquaintance on the street; I cannot remember his name. "Ah! It's stupid," I tell myself: "You know him so well . . . Come on, what's his name?" A few hours later, talking about something else: "That's it, it's coming back to me: Mr. X!"(figure 2.15). It is a benign omission, very common, especially among people who are depressed or overworked. The name recovered, the word found; so, it was present in my brain. For the right word to have been found, and not another, a mental process must have taken place in my brain—without my thinking about it, without my realizing it.

Between the time the name was lost and found again, my brain has been working without realizing it. To obtain such a precise answer to such a precise question after a considerable period of time presupposes that there has been an intellectual journey of which I was not aware. What could this be, if not subconsciousness? The problem remains how to explain this sudden "unintentional" return to the consciousness of a mental process that has been produced subconsciously.

2.15 Remembering a name.

The subconscious nature of the recall mechanism is, in fact, difficult to demonstrate, because it is necessary to ensure that between the initial memory gap and the unexpected recall, the subject has not thought about the missing information. The existence of such a subconscious reminder has been well demonstrated by studying patients who have lost conscious memory (pure amnestic syndrome) but who, after several attempts, can report forgotten memories, thus indicating that the progressive enrichment of these memories has been carried out subconsciously (box 2.2).

BOX 2.2: A SUBCONSCIOUS RECALL

A forty-five-year-old woman lost her memory following a brain hemorrhage that had totally destroyed the brain structures involved in the control of memory (i.e., the internal part of the temporal lobe of both hemispheres). Although alive, she was severely disabled in immediately forgetting what had happened or had been said to her moments before. She had no conscious memories since the time of her accident. Taking advantage of this striking, although very disabling disability, François Lhermitte showed the patient a detailed schematic drawing of a district in Paris where she had lived in the past. Over five successive days, he asked her to describe the different parts of that district. After each day, the patient was unable to remember the previous session, indicating that she was not aware of being asked the same questions regarding a previously well-known area and did not even recognize the examiner. The first day, the answers were very poor: the drawing was almost empty. But during the following days, the responses were progressively improved to the point that, on the fifth day, she was able to provide all the details pertaining to that part of the Paris map.

In this case report, despite the massive loss of conscious memory, the items that had been memorized before the accident were not lost. They were only not accessible. Between each session, the patient's brain was performing an automatic mental processing without her being aware. Although the neuronal circuits which elicit the conscious recall of memories were wiped away, other neuronal pathways remained intact, thus allowing complete subconscious recall of memories.[1]

---

1. François Lhermitte and Michel Serdaru, "Unconscious Processing Memory Recall: A Study of Three Amnestic Patients," *Cortex* 30 (1993): 35–42, doi: 10.1016/s0010-9452(13)80209-2.

"Hysteria": Psychogenic Diseases

Some patients present a set of symptoms that resemble a neurological disease but without any detectable damage to the nervous system. Paralysis, blindness, contracture, and abnormal movement are examples, but all without an organic substrate. This is what was once called in the past "hysteria." Psychogenic diseases must be distinguished from simulation, in which the patient consciously tries to benefit. In contrast, in psychogenic diseases, the patient is convinced that he or she is sick and is not aware that there is no physical basis for the disease. The patient thinks that he or she cannot act (paralysis) or perceive (blindness, anesthesia), when, except for the psychological barriers, he or she is not impaired (figure 2.16).

2.16 Psychogenic blindness.

The symptoms are identical to the spontaneous symptoms observed during nervous system diseases, but "the nerve cells remain inert as if they were destroyed," said Jean Martin Charcot, who considered that psychology was "nothing but the physiology of the superior, noble parts of the brain."[5] It is as though the responsible neural circuits have lost their function while remaining intact. The patient is normal, mental functions are intact, but he or she presents symptoms that he or she invents without knowledge, without awareness—in other words, subconsciously. Psychogenic symptoms can disappear if the psychological blockage is broken, because the nervous system is intact. In practice, the situation is reversible, but not always and for reasons that are not entirely clear.

Intuition: Scientific Discovery

A third example of this type of subconsciousness is the period immediately preceding a scientific discovery. The discovery is not the result of a series of rule-guided symbolic manipulations, proceeding in the manner of a computer testing all possible combinations to win a game of chess. Nor is it a rational language. Albert Einstein said, "I very rarely think in words at all. A thought comes, and I may try to express it in words afterwards."[6] It is akin to the feeling when you meet someone for the first time; you do not know them, yet you immediately feel sympathy or antipathy.

Sensitivity is also involved in scientific discovery; it involves elegance, the beauty of a thought pattern, the joy of having found the solution. Rather than being the result of a chain of conscious reasoning, the creative process is essentially subconscious. Indeed, when we want to discover, we start by forcing ourselves to think, but the explanation rarely comes in full consciousness. Most often, after searching in vain, the solution suddenly emerges. It was Henri Poincaré who suddenly found the solution to his famous theorem by getting on the bus in the small town of Coutances in Normandy. He describes the experience: "At the moment when I put my foot on the step the idea came to me, without anything in my former thoughts seeming to have paved the way for it." In reflecting on his experience, Poincaré describes the appearance of such "sudden illumination" (of which he has multiple examples) as a "manifest sign of long, unconscious prior work."[7]

Between the phase of primordial questioning and the moment of discovery, there is intellectual work going on to which we do not pay attention, of which we are not even aware (figure 2.17). It is like dreaming: myriad subliminal thoughts collide in the brain, one of which eventually exceeds the threshold that brings it to consciousness. In other words, discovery is not, in general, the result of a series of conscious reasonings; it is an essentially subconscious creative process. The hypothetic mechanism of dreaming, a sort of chaotic movement of ideas and images like the Brownian

2.17 Subconscious thoughts awaiting discovery.

movements seen with atoms, might—why not?—be applied to the unbelievably complex cybernetics of thoughts and emotions appearing in the state of subconsciousness.

In these examples, the subconscious processing of information that has been either the subject of significant intellectual effort or triggered by a particular emotional situation makes it possible for me to recall a name that had escaped, for symptoms to appear without an organic substrate, and for a potentially brilliant discovery to occur. We have moved from a subconscious state to a conscious state. On the other hand, information that has been consciously stored has also been conveyed to the depths of subconsciousness. The working hypothesis is therefore that these different states of consciousness are linked in both directions.

The purpose of these introspective deductions is not to add one more hypothesis to existing theories of brain functioning. Instead, the aim is to reflect on the possibility of a physiological basis for these different forms of consciousness; in particular, to reinforce the concept of subconsciousness and to discover its anatomo-physiological substratum.

3

A Physiological Base for the Subconscious in the Brain

I n simple terms, humans can adopt two distinct, albeit linked, types of behavior: the design of a new behavior to adapt to an unexpected or difficult situation (consciousness); and the expression of automated motor nonconscious programs (subconsciousness modulated by consciousness). One way to approach the essence of these two major mental functions is to discover the mechanisms responsible for them within the nervous system. The brain is obviously in the front line, for we know that it produces our behaviors and has the exceptional quality of governing both body and brain. The brain is indeed is a masterpiece.

The brain is *expensive*. Weighing nearly three pounds, or about 2 percent of the body's weight, it consumes 20 percent of the body's energy.

The brain *grows*. At birth, its weight is just under 25 percent of that of adults, and it continues to increase until late adolescence, when it weighs 1,300 grams for women (1,350 grams for men, because they are taller).

The brain is *complex*. It contains about 85 billion neurons, which constitute innumerable circuits in the brain, at least as many nonneuronal cells (called glial cells, from the Latin word for "glue") that interact with neurons and ensure their survival, and nearly 600 kilometers of blood capillaries.

The brain is *moving*. It is constantly modified according to what we are learning, allowing the rearrangement of nerve connections.

The brain is *stubborn*. Neurons do not multiply, which means that, in a centenarian, the neurons in the brain have lived for a century.

The brain is *intelligent*. It listens, decides, and acts. It is therefore a machine with complexity beyond our imagination, although we are beginning to understand its foundations.

Within the brain, this gelatinous mass located under the skull, what might be the brain structures playing a role in subconsciousness (figure 3.1)? There is every reason to exclude the "executors"—the spinal cord and peripheral nerves—which transmit the messages back and forth to the brain, as well as the brainstem, which controls vegetative functions, and the hypothalamus, which controls the secretion of hormones necessary for the functioning of our viscera.

To briefly and simply illustrate the neurophysiological concept (without detailing the role of several important brain areas located in the depths of the brain that might contribute in a limited way to the function of subconsciousness), areas of activation include: the amygdala, the "alarm," which is rapidly activated each time a subject perceives a strong unpleasant emotion such as fear; the cerebellum, the "adjuster" involved in coordinating movement and balance; and the thalamus, the "relay station" for messages intended for the cerebral cortex. However, the first-line candidates for generating behavior are essentially the cerebral cortex, which is phylogenetically recent, and the older basal ganglia. With this deliberately oversimplified dichotomy, how can we go about identifying the neurophysiological correlate of subconsciousness?

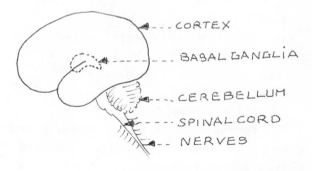

3.1 The brain and its executors.

An enlightening way of distinguishing between nonautomatic conscious behaviors and automatic subconscious behaviors is to refer to pathology. For example, in Alzheimer's disease, the patient has intellectual disorders (memory, language, perceptions) resulting from dysfunction of different areas in the cerebral cortex, while automatic behaviors (walking, eating, exhibiting elementary courtesy) that depend on the proper functioning of the basal ganglia remain unaffected until well into disease progression (box 3.1).

BOX 3.1: A SUSPICION OF ALZHEIMER'S DISEASE

Mr. X, a new patient accompanied by his wife, has come for a consultation. He appears to be in his mid-70s. He greets me courteously, gets up from his chair, walks to my office, and sits down. Nothing special about his movements, which are performed normally.

"What can I do for you?"

He turns his head toward his wife as if he is asking her to answer and says: "I don't know."

His wife: "Well, Doctor, he is losing his memory."

He remains silent.

"Can you give me an example?"

"Oh, yes, several! For instance, he is always asking the same questions. If I ask him to do some shopping and give him a piece of paper with the list of what to buy, when he comes back, he's forgotten to buy most of the food and has lost the paper. He is even forgetting the names of his grandchildren. But you know what is strange, he remembers things that happened in the past, like our wedding."

"Is he oriented to time and place?"

"You are right, sometimes I have the impression that he does not know where he is. Anyway, he is totally lost with the dates."

> "Otherwise, his general behavior is normal?"
>
> "Yes. He sleeps normally, he eats normally, he walks normally."
>
> "Mr. X, are you concerned?"
>
> "Not really."
>
> "He is not, but I am! He is a person who is usually so calm and quiet, but now sometimes gets angry for no reason. The other day, he accused me of taking money from his wallet!"
>
> Mr. X already has severe memory loss of recent memory; but his former memory seems relatively preserved (suggesting a dysfunction of the memory circuit involved in recent memory, in particular the internal temporal cortex), indicating that at least part of his intellectual nonautomatic behavior is altered (consciousness). He does not seem to be aware (anosognosia) or to pay attention (anosodiaphoria; consciousness). Moreover, he is disoriented with respect to time and space and exhibits unusual psychiatric reactions. In contrast his automatic motor behavior (gait, etc.) are spared (subconsciousness), suggesting that the basal ganglia are preserved.

The situation of the Parkinson's patient is the opposite: automatic behaviors such as walking, brushing teeth, and writing are altered due to the dysfunction of the basal ganglia, while cognitive functions (mainly managed by the cerebral cortex) are long preserved (box 3.2).

BOX 3.2: IS IT PARKINSON'S DISEASE?

Mr. Y, sixty years old, walks slowly, his feet scrubbing lightly on the floor. He is slightly bent down, his right arm does not swing normally, his face is relatively immobile, and his voice is monotonous.

"What bothers you the most, sir?"

"My writing is smaller and smaller. I have a problem slipping on my shirt. When my wife and I take our walk, I noticed that I tend to remain behind her. Sometimes, when my right hand is at rest, it starts to tremble."

"And you, madam, what do you think?"

"What my husband says is absolutely correct. He was so active in the past, and now he does not do much. I wonder whether he is not depressed."

"Tell me, doctor, will I end my life in a wheelchair?"

"There is no reason for that! You must understand that what you are suffering from can be efficiently treated."

This man is more and more disabled by slowly progressing motor symptoms predominant on one side of the body. The loss of his automatic movements (decreased arm swing, slowness of gait, amimia, micrographia) suggests Parkinson's disease. This is in favor of a dysfunction of the basal ganglia. He is probably depressed, but his intellectual abilities are preserved, suggesting that the cerebral cortex is intact. He is aware of his difficulties but is consciously worried about the future. Fortunately, he will benefit from an efficacious treatment, namely a replacement therapy to reestablish normal dopamine transmission in his brain.

These two pathologies are somehow mirrored, which suggests, but does not demonstrate, that the cerebral cortex plays a predominant role in nonautomatic behaviors, and conversely, that basal ganglia dominate in automatic behaviors (figure 3.2).

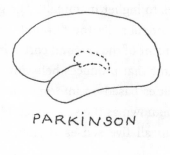

PARKINSON

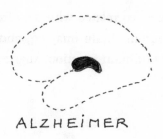

ALZHEIMER

3.2 Two "mirror" neurological disorders, Alzheimer's disease and Parkinson's disease. In Parkinson's disease, at least at the onset of the affliction, the cerebral cortex is spared, while the basal ganglia (in the center of the brain) do not function properly (dotted line). In Alzheimer's disease, the cerebral cortex (dotted line) malfunctions, while the basal ganglia (in black) are intact.

The Cerebral Cortex, the Main Actor in Nonautomatic Behaviors

Developed mainly in so-called higher animals, the cerebral cortex consists of a layer of nerve cells located in the peripheral part of the brain (figure 3.3). It develops in a dazzling way from reptiles to higher mammals. The great winner of this growth is undoubtedly the "frontal" anterior cortex, occupying 30 percent of the cerebral cortex in humans. It is the brain's effector that produces behavior, that translates into motor activities based on information received by the posterior cortex, known as the sensory cortex, and receives information from all five senses (vision, touch, hearing, smell, taste).

Clinical studies, correlated with data from the electrical recording of neurons, brain imaging, and different mathematical models of brain function, suggest that our cortex

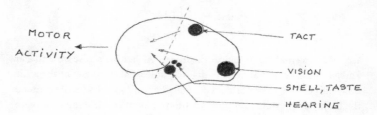

3.3 The cerebral cortex: the posterior perceives, the anterior acts. All perceptions (tact, vision, smell, taste, hearing) are integrated into the posterior part of the brain that transmits information to the frontal anterior part of the brain, that determines our motor behavior.

contributes predominantly to consciousness and metaconsciousness skills. These behaviors are most often produced from "external" stimuli from the environment, but the cerebral cortex can be brought into play from an "internal" decision that implies that we consciously anticipate performing a behavior.

These observations are consistent with pathology data that show that various perceptual functions are impaired in the case of posterior cerebral cortex injury; for example, neglect of the entire contralateral hemispace in the case of selective right parietal cortex injury (figure 3.4) or total

3.4 Oblivious of the left space, right cortical lesion. Following the obliteration of a cerebral artery in the region of the parietal cortex of the right cerebral hemisphere, this patient is ignoring the left side of his body and its surrounding space. He has no perception of anything happening on his left. It is as though he has forgotten its very existence, and he is often not even aware of not being aware.

blindness in the case of posterior visual cerebral cortex injury (figure 2.12). In all these cases, patients lose awareness of their lack of perception, but they can think and act in a nonautomatic way (except, of course, in the perceptual field of their disability), which means that their consciousness is also preserved. The same is true of routine, subconscious activities, suggesting that they do not depend directly on the functioning of this part of the cerebral cortex. In short, lesions in the posterior (perceptual) cerebral cortex result in a loss of the specific functions of each of these parts (vision if occipital cortex lesion; language if left temporal cortex lesion in a right-handed person, etc.) but the consciousness and subconsciousness functions have been left intact, at least when the lesion is restricted to the cerebral cortex.

The situation is more complex when the frontal cortex is destroyed. In the most severe cases, such as frontal dementia, for example, not only do patients lose awareness of their disability, but they also lose the ability to reason and plan an action. On the other hand, here again, the motor, intellectual, and emotional routines are preserved. Sometimes they are even exaggerated, as if they were liberated: the patient makes inappropriate or even incoherent gestures, no longer controls impulses (bulimia, exhibitionism), and becomes indifferent to the outcome of previously valued actions or goals (box 3.3 and figure 3.5). In short, automatic activities are retained, but they are often poorly controlled, because the frontal cortex no longer does its job as a conductor. Nonautomatic behavior, depending upon the functioning of the frontal cortex (conscious) tends to become automatic (subconscious) as a result of the liberation of the basal ganglia, which are now not under the control of the frontal cortex.

BOX 3.3: THE NAKED WALKER—DYSFUNCTION OF THE FRONTAL CORTEX

For at least a year, this level-headed accountant had been acting strangely. Recently, his wife had received a telephone call from his closest colleague at work saying that her husband, known to be particularly meticulous, was no longer taking care of necessary paperwork or responding to the mail. The call was not entirely a surprise, as she was already worried. He seemed to have lost interest in everything, he looked indifferent, and would have sudden bursts of anger for no reason. She said that he was now spending most of his time, immobile, in front of the TV set, except for those times when he abruptly decided to consume the entire contents of the refrigerator (bulimia). Distressingly, he was taken by police to the emergency room after he was found walking naked in the street (disinhibition). When the policeman questioned him as to why he was naked, he said: "I don't know. I don't mind" (anosognosia, anosodiaphora). During the consultation, gait and motricity were without particularity, memory seemed preserved, as were all perception abilities, and the rest of the examination was normal. Faced with this odd psychiatric picture, a dysfunction of the frontal lobe was suspected, and the brain scanner indeed showed a marked bilateral atrophy of the frontal lobe in favor of a "frontal dementia."

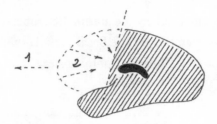

3.5 Dysfunction of the frontal cortex. Although the rest of the brain is normal (hatched), the dysfunctional frontal cortex (dotted line) is no longer able to correctly express nonautomatic, conscious (1) behavior and no longer exercises inhibitory control (2) over the basal ganglia (black), so conscious, cortical behavior tends to become automatic, subconscious as a result of the continuing activity of the basal ganglia.

In a word, when a dysfunction selectively affects a part of the cerebral cortex, the faculties of consciousness and meta-consciousness may be altered, but this is not the case for sub-consciousness, which is left intact, quite simply because the brain structures that, for the most part, generate and control subconscious automatic behaviors remain unharmed.

The Basal Ganglia, the Centerpiece of Subconsciousness

What would our existence be like if we had to constantly think about our actions (walking, cycling, directing the movements of the tongue and mouth to express ourselves, driving our car, etc.)? It is not possible to consciously think about our every move, fraction of a second by fraction of a second. Fortunately, this "subconscious" work, boring and laborious, exhausting due to its permanence, is essentially carried out by the basal ganglia. And yet, there are reasons as to why little attention has been paid to these structures.

Located deep within the center of the brain, at the base of the two cerebral hemispheres ("the dark basement of the brain," as described by Kinnier-Wilson), their volume is ridiculously small compared to the total weight of the brain: 30 grams, the equivalent of two 2.5 centimeter by 2.5 centimeter cubes, or almost one-fiftieth of the brain's weight (figure 3.6)! These small balls stuffed with neurons (a few hundred thousand) receive messages from the cerebral cortex (20 billion neurons), which results in a spectacular

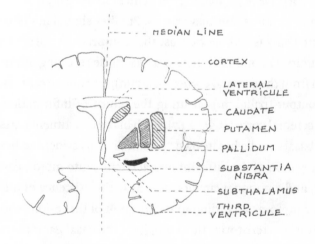

3.6 The basal ganglia: a small brain on their own. Located in the center of the brain, the basal ganglia comprise several substructures, each of which has a more specific role in controlling automatic subconscious behavior. Schematically, lesions in each of these areas lead to a particular clinical picture: caudate, chorea; putamen and pallidum (with two external and internal parts), dystonia; substantia nigra, Parkinson's syndrome; subthalamus, hemiballismus.

compression of cortical information toward the basal ganglia (a ratio of almost 1,000:1)! The smallness of basal ganglia cells in relation to the cerebral cortex paradoxically becomes a therapeutic advantage by allowing the information from the cerebral cortex to be selectively modulated within a limited volume of brain tissue.

Disregarded due to their subordinate role in the execution of behavior, abandoned in favor of the cerebral cortex (considered as the noble brain regions par excellence), distorted because they are supposed to be involved only in the narrow management of automatic (e.g., swinging of the arm to walk) and repetitive (e.g., walking) movements, *basal ganglia are in fact the true touchstone of subconsciousness*. But they should not be considered only as a whole, because they comprise several separate subunits. On one side, there is a receptacle receiving information from the cerebral cortex (striatum), and on the other side, an output (pallidum) ensuring the return of information to the cerebral cortex (figure 3.7). Each of the constituent parts of the basal ganglia has its own "personality," as evidenced by the segregation of the different neural circuits into motor, cognitive, and emotional circuits, as well as the existence of different symptoms according to the location of the lesions within the various territories that constitute the basal ganglia.

These brain structures appeared early in evolution, well before the appearance of the cerebral cortex, the latter allowing for the deployment of cognitive functions at their peak. But, to momentarily anthropomorphize, if we are old and of limited volume, does it mean that we are less valuable? In any case, the basal ganglia cannot be criticized for being

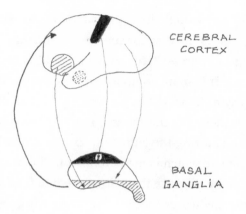

3.7 The basal ganglia–cortex couple. The different parts of the cerebral, motor (in black), intellectual (associative, in white), and emotional (limbic, hatched) cortices are projecting toward the motor, associative, and limbic parts of the basal ganglia, respectively.

fragile: they have allowed animals (such as insects, fishes), which have a restricted cerebral cortex, to survive for hundreds of millions of years. Widely predominant among reptiles and birds, they are the main actors in automatic behaviors, thanks to their role in the execution, selection, and learning of our motor skills but also in our intellect and emotions.

The Role of the Basal Ganglia

1. *The basal ganglia have a role in the automatic execution of movements.*

They perceive information from the environment (via the cerebral cortex), store it, and initiate motor behavior autonomously.[1] However, each component of these automated movements includes a collection of independent motor sequences that ensures the coordination of the entire body in action. This is the case for simple motor behaviors, such as walking or running (arm swaying, trunk and shoulder oscillations). It is more complicated for complex behaviors such as playing tennis (box 3.4 and figure 3.8) or speech (micromovements of the mouth, tongue, and particularly the larynx), allowing the automatic expression of language (the semantic of which is not automatic). The basal ganglia thus have an intelligence that allows them to fend for themselves while the "driver," the cerebral cortex, is in freewheeling mode. However, even if "ils ont la bride sur sur le cou" (meaning the ability to act freely), they remain under the control of the cerebral cortex. If they can be autonomous, it is because, on their own, they are able to perceive information from the environment, store it, and initiate the resulting driving behavior, demonstrating that their role is not only to carry out the orders of the cerebral cortex.

BOX 3.4: WHY ARE PROFESSIONAL TENNIS PLAYERS UNBELIEVABLE CHAMPIONS?

When a tennis player receives the service (200 kilometers per hour) from his opponent, he only becomes conscious of the trajectory of the ball when it has cleared the net, that is, after 285 milliseconds.[1] This means that the player has already started a series of concomitant and sequential movements, long before being aware that the ball is heading toward him. He does not have the time to think about the complex movements he must perform (as if it were a nonautomatic conscious motor behavior). In fact, his brain has predicted the different motions necessary to return the ball. These incredibly rapid movements are accomplished without the player being aware, that is, subconsciously. If the ball does not arrive where he had predicted, he will miss it. If he has correctly anticipated the trajectory of the ball, he might then consciously catch his opponent by sending the ball back out of his reach.

1. Rita Carter, *The Human Brain Book* (New York: Dorling Kindersley, 2009).

3.8 The tennis champion. To become such a superb tennis player, this champion must have learned to play tennis many years before. At first, when he started to play, his forward and backhand shots were obviously clumsy. He was consciously learning by trial and error, thanks to the good functioning of his cerebral cortex. After months and years of training, however, he developed all the movements needed to control the ball, including the service, overhead, and drop shots, without his even being conscious (i.e., subconsciously.) From now on, his cerebral cortex is focusing on beating the adversary, whereas his basal ganglia are taking care of all his overlearned effective automatic movements, as demonstrated experimentally in rodents[1] and using neuroimaging in humans.[2]

1. Ann Graybiel, "The Basal Ganglia: Learning New Tricks and Loving It," *Current Opinion Neurobiology* 15 (2005): 638–44;

2. Stephane Lehéricy et al., "Distinct Basal Ganglia Territories Are Engaged in Early and Advanced Motor Sequence Learning," *Proceedings of the National Academy of Sciences USA* 102 (2005):12566–71.

2. *The basal ganglia are able to sort the countless pieces of information from the cerebral cortex, keeping those that are most relevant to allow agile behavior and permanent adaptation to the environmental situation.*

They thus facilitate movement by modulating its amplitude and velocity, and by eliminating inappropriate or unwanted movements (figure 3.9).[2] This is not an easy task, because any automatic movement, even the most elementary, includes a constellation of bits of routine movements.

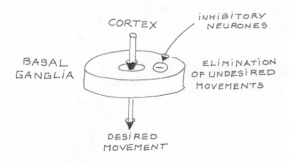

3.9 Basal ganglia can eliminate unwanted movements to produce the desired movement. Messages from the cerebral cortex are cleared and then sent back to the frontal cortex, which finalizes the movement. To accomplish this task, the basal ganglia eliminate unwanted movements, thanks to inhibitory neural influences, which allows the desired movement to be achieved.

3. *The basal ganglia serve as a wonderful learning tool.*

When a baby learns to walk or when a child first learns to ride a bicycle, the actions are done awkwardly. Missing, one tries again; if movements are poorly executed, one improves by successive trial and error until the maneuvers are smoothed out, adapted, graceful, automated. With training, movements are eventually performed automatically without thinking about them, without question. In this way, the basal ganglia acquire a personal know-how that allows them to transform new information into routine that they store. Once the motor behavior is acquired within the basal ganglia, our actions become autonomous with respect to our daily motor activities.[3] These memories are so well imprinted into the basal ganglia that it becomes almost impossible to get rid of them. This explains why complex learning, such as skiing or waltzing, is memorized, to the point of allowing us to ski or dance after long periods of interruption.

4. *The basal ganglia have an intellectual and emotional function.*

They play a crucial role in the control of emotions, as has been shown: on the one hand, during selective dysfunction of these structures in patients[4] or during manipulation of the "emotional" territories of these structures, either experimentally[5] or in patients (box 3.5); on the other hand, with the help of neuroimaging, where one can show that it is the "emotional" territories of the basal ganglia that are activated preferentially during even subtle emotions such as romantic love.[6] What is true for humans is true for all evolved animals.

A sixty-five-year-old woman with a severe thirty-year history of
Parkinson's disease underwent a neurosurgical operation (bilateral
implantation of electrodes in the region of the subthalamic nucleus,
which is part of the basal ganglia). Instantly, she was remarkably
improved. To identify the optimal site of stimulation, during the
postoperative period, according to a standardized protocol, we
began to investigate other stimulation sites in this region.

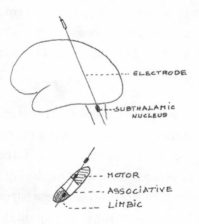

Suddenly, stimulation of a site that was not the usual therapeutic target induced a dramatic clinical picture. The patient started
to cry, expressing sadness, uselessness, and guilt: "I no longer
wish to live . . . I'm fed up with life . . . It's no use . . . I want to
hide in a corner . . . Why am I bothering you?" During this short
period, she was alert and recalled the entire episode. Immediately,
as soon as the stimulation through the electrode that had induced
the unexpected severe side effect was interrupted, this behavior

disappeared; the patient was released from the disturbing symptoms that were absent when the original therapeutic target was
stimulated. Anatomical localization of the electrode inducing this
transient acute depression showed that the site of stimulation
within the subthalamic area was selectively activating a neuronal
circuit implicated in the production of emotions (as shown by pos-
itron emission tomography).

This case report shows that the modulation of a restricted
territory of the basal ganglia is able to induce a complex psychiatric behavior such as depression.

Source: Boulos Bejjani et al., "Transient Acute Depression Induced by High Frequency
Deep Brain Stimulation," New England Journal of Medicine 340 (1999): 1476–80, doi:
10.1056/NEJM199905133401905.

After all, it is not surprising that the basal ganglia have
a role in the management of emotions, given that neurons
from the cerebral cortex, destined projecting to the basal
ganglia are divided into three separate neuronal contingents
that underlie the transfer of motor, cognitive, and emotional
activities, respectively (figure 3.9). It is therefore no wonder that these structures have not only a driving motor role
but also an emotional one, even if these territories are not
"watertight" and communicate with each other through neuronal collaterals impinging on neighboring territories.

Moreover, the basal ganglia likely contribute to the fusion of intellectual and emotional messages to produce a synthetic motor behavior,
the ones we see and hear.[7] Thus, the intellectual and emotional
messages received in the different parts of the cerebral cortex
(whose surfaces are extensive) are conveyed to the basal

ganglia (which have limited volume) where they would be synthesized to produce a movement that expresses the behavior. In other words, motor behavior, the behavior we see (gestures) and hear (speech), is only the expression of our ideas and emotions. In this sense, the basal ganglia that ultimately express motor behavior actually reveal the depths of each person's personality, the one that is expressed without our knowledge, but that does not escape a discerning observer.

5. *The basal ganglia are faithful collaborators of the cerebral cortex.*

They are not alone, isolated and cut off from the rest of the brain, as they involve the cerebral cortex every time they are activated, just as they are activated every time the cerebral cortex drives deliberate behavior. This is easy, because they work in a loop with the cerebral cortex (figure 3.7). For each behavior, there is a real reprogramming of the motor loop between the cerebral cortex and the basal ganglia. This greatly benefits the cerebral cortex: freed from the task of directing what are now automatic behavioral activities, it can engage in tasks requiring new adaptation to the environment. Most often, however, as soon as the cerebral cortex disengages, the basal ganglia become independent, because they have an intelligence that allows them to fend for themselves, in short, to play their role as automatic pilots. One might say that this is a caricatured reductionist perspective. The physiological reality is much more complex—which will undoubtedly prove to be true in the future—so this oversimplification must be tempered, because any involvement of one of these brain structures actually engages and activates the whole brain.

4

The Diseased Subconscious and Implications for the Future

Theory is good, but it doesn't prevent things from existing.

—JEAN-MARTIN CHARCOT

To approach the subject of subconsciousness under normal and pathological conditions, my argument has been necessarily equivocal, because it is addressed both to experts in the field who would certainly detect banalities in it and to science popularizers who would undoubtedly point out the unclear and obscure parts. Whatever is said about it, the aim here is to draw attention to the fact that automatic behaviors that constantly ensure our survival can be the root of many neuropsychiatric disorders. Since these behaviors are essentially managed by the oldest structures of the brain makes the latter potentially overlooked therapeutic targets.

When my movements are inappropriate, my thoughts incorrect, and my emotions harmful, they are stronger than my ability to resist them. However, as discussed in previous chapters, these symptoms are not observed when only the cerebral cortex is the site of dysfunctions (frontal syndrome

in the anterior part of the brain, elementary or complex difficulties of perception in the posterior part of the brain).
Instead, these symptoms are present when the basal ganglia
do not function efficiently, resulting in involuntary abnormal movements, thoughts, and emotions. The basal ganglia
could, therefore, represent a potential target for treatment
of a range of neurological and psychiatric diseases, using
pharmacological or neurosurgical tools.

First, it is possible to improve symptoms by selectively
reactivating deficient neuronal circuits or blocking overactive neuronal circuits within the basal ganglia, which are
known to be enriched in neurotransmitters such as dopamine, serotonin, acetylcholine, and gamma-aminobutyric
acid (GABA). Hence, the use of dopaminergic agonists (and
primarily the dopamine precursor, L-DOPA) allows resto-
ration of the deficient dopaminergic transmission that is
characteristic of Parkinson's disease (box 3.2). Conversely,
neuroleptics, dopamine receptor blockers, are prescribed
to reduce the hyperfunctioning of these same dopamine
neurons to erase the delusional episodes that characterize
schizophrenia (figure 4.1).

Another example is depression, a condition that is largely
ameliorated by the use of antidepressants, which are known
to increase serotonin concentrations. Although there can
be some prescription abuse of these types of drugs, there is
a tendency to overlook the fact that the use of antidepressants has transformed the long-term prognosis of depres-
sive illness. Examples could be multiplied, with the use of

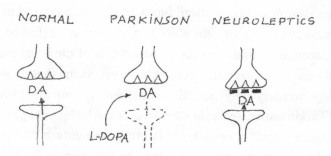

4.1 How is dopaminergic neurotransmission activated or inhibited? In normal conditions (*left*), the neurotransmitter dopamine is released by the dopaminergic neuron into the cleft (synapse) to stimulate the dopaminergic receptors (in white) located downstream on the next neuron, thus eliciting a normal dopaminergic transmission. In Parkinson's disease, the release of dopamine is reduced as a consequence of the degeneration of dopaminergic neurons, a characteristic of the disorder. Dopaminergic receptors on the following neuron are less or no longer active, resulting in reduced dopaminergic transmission at the origin of symptoms. The treatment consists of replacing the deficient dopamine with newly formed dopamine from the administration of the precursor of dopamine, L-DOPA. In patients with schizophrenia, overactive dopaminergic transmission, which is considered to contribute to the appearance of delirium, is reduced by the administration of neuroleptics, known to block the dopamine receptors, thus relieving the most severe symptoms in patients.

anticholinergics in certain types of tremors or in dystonia or of benzodiazepines (GABA activators), which consistently overcome the most severe anxiety attacks.

Second, it is even possible to modulate neuronal activity within the basal ganglia using neurosurgical procedures. With the disadvantage of surgical risks associated with neurosurgical intervention, but with the advantage of precision and selectivity, the placement of electrodes (linked to a pacemaker) in one of the substructures of the basal ganglia can give interesting results in severe forms that do not respond adequately to medical treatment: certain rare forms of Parkinson's disease, in which where the lesions are purely dopaminergic (figure 4.2);[1] twisting movements and abnormal postures observed in dystonia, a frighteningly disabling disease that had no treatment in the past;[2] and Tourette's syndrome, characterized in its most serious form by violent tics, aberrant vocalizations, and self-mutilations.[3] Current scientific advances in this field are such that it is even possible to consider the symptomatic treatment of psychiatric disorders such as certain severe obsessive-compulsive disorders that are beyond any other medical therapeutic help (box 4.1).

For the future, the therapeutic prospects are somewhat encouraging, not only with the development of drugs to prevent the pathological process of these diseases, but also with other innovative technical processes. Although it is true that the treatment of Parkinson's disease by transplanting embryonic dopamine neurons has been abandoned (due to therapeutic insufficiency and the severe side effects that may result), new therapeutic possibilities are opening up in the near future thanks to the development of ultrasound neurosurgery (which precludes penetration through the skull), gene therapy (e.g., by suppressing a mutation), targeting drugs in a specific disease area (e.g., with monoclonal antibodies), and providing trophic factors that allow neuronal growth to occur.

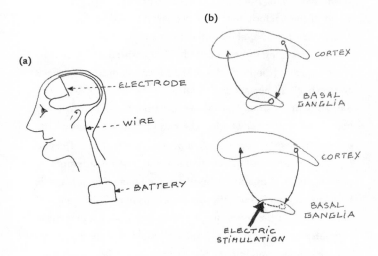

4.2 Principle of treatment of certain forms of Parkinson's disease using electrodes implanted in the brain. (a) The electrode, connected to a battery (pacemaker) through a wire established under the skin, is introduced in a given substructure of the basal ganglia. (b) In its normal state (top), the cerebral cortex sends the basal ganglia the information it has received. The basal ganglia return this refined information to the cortex, producing appropriate behavior. In case of dysfunction of the basal ganglia (for example, in Parkinson's disease), physiological messages are no longer sent back to the cortex (dotted line), which determines the symptoms. In select patients, it is possible to reactivate neurons projecting to the cortex by using electrodes to stimulate healthy neurons located downstream of diseased neurons (electrical stimulation).

BOX 4.1: THE MADNESS OF DOUBT

Miss Y is a thirty-five-year-old patient with a long history of obsessive-compulsive disorder. At the age of twelve, she began washing her hands several dozen times per day, saying she was afraid of the bacteria that were present on each "dirty object." After a few months, she started to spend several hours every evening arranging toiletry objects in her bathroom. Her performance at school diminished, and she became compelled to check and recheck all her actions, from what she wrote to what she did. Sometimes when leaving her apartment, she had to verify more than ten times, whether she had forgotten to shut off the gas or whether the front door had been correctly closed. Despite repeated sessions of psychotherapy and several attempts at pharmacological treatment, she continued spending at least six hours per day performing these bizarre behaviors. This intelligent patient was all the more depressed by being aware that these compulsions were senseless, but she could not control them. Her life became a nightmare. She had repeatedly been fired from work, her husband had left her, and her children and friends considered her "crazy" and did not want to see her anymore. She felt she had nothing to live for and wanted to kill herself. To relieve her suffering, a different approach was implemented. The bilateral implantation of electrodes in the "emotional" territory of the basal ganglia (subthalamic nucleus) resulted in a 90 percent improvement of her compulsions (washing, ordering, checking) and 60 percent improvement of her obsessions (dirt, untidiness, doubt). After the intervention, she was able to resume working again and start a new and productive life.[1]

The neurosurgical permanent modulation of the neuronal circuits controlling emotions in the basal ganglia made it possible to eliminate these "abnormal involuntary emotions" (emotional subconsciousness), namely the severe obsessions that the patient tried to compensate for by using various types of compulsions.

1. Luc Mallet, Mircea Polosan, et al., "Subthalamic Nucleus Stimulation in Severe Obsessive-Compulsive Disorder," *New England Journal of Medicine* (2008): 2121–34.

Specifically, at a time when most investigators are trying to find therapeutic approaches that target the most recent and "noble" brain areas (the frontal cerebral cortex, the brain conductor; the hippocampus, the "tollgate" of memory"; the amygdala, the touchstone of harmful emotions), the purpose here is to draw attention to another potential therapeutic target—the basal ganglia, for three reasons:

The first is anatomical: the extreme convergence of the major neuronal pathways from the different areas of the cerebral cortex, which include perfectly segregated neural circuits (motor, associative, and limbic, supporting–respectively– motor, intellectual, and emotional information), are concentrated in small volumes easily accessible for therapeutic targeting (figure 3.9)

The second is physiological: the basal ganglia play an indisputable role in the genesis and control of the subconscious function, the importance of which, as we have seen, ensures our essential daily behaviors (autopilot) and even our survival.

BOX 4.2: "NOTHING HAPPENS IN MY HEAD": MENTAL EMPTINESS RESULTING FROM EXTENSIVE DESTRUCTION OF THE BASAL GANGLIA

Mr. Z suffered a severe anoxic coma following a wasp bite. He woke up twenty-four hours later exhibiting abnormal involuntary movements of the forelimbs mimicking chorea (from the Greek word for "dance"). When examined several years later, he exhibited strange behavior. For hours on end, he remained inert, doing nothing. If someone asked why he was not moving, he replied, "Nothing happens in my head. It's totally empty." When questioned how that could be, as it is impossible not to think, he replied: "It's impossible for you, Doctor, but not for me." He was not bored, he did not suffer, he was not depressed. As remote as he seemed when alone, it was astounding that when stimulated by someone else, he immediately retrieved his previous intellectual abilities until left by himself again he retreated to his prostrate state. When by himself, he was completely inactive, except for repeatedly counting: "one, two, three, . . ." until he reached twelve, and then he would start again from zero. He explained, "There is an internal force stronger than I am that is pushing me, and I am unable to resist."

In sum, a strange apragmatism and a sort of peculiar obsessive-compulsive disorder. The brain scan showed a complete destruction of the basal ganglia with the cerebral cortex left intact.

Such a clinical picture was named "loss of psychic auto-activation" by Dominique Laplane,[1] meaning that the patient needs a stimulation from the outside world to be able to think and to feel. It is hypothesized that the massive bilateral lesion of the basal ganglia is no longer able to activate the frontal cortex, which remains intact.

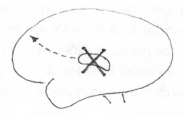

In this type of patient, consciousness is preserved, in line with the preservation of the cerebral cortex, provided the patient is stimulated; but subconsciousness is dysfunctional, as suggested by the bizarre, repetitive, compulsive behavior, compatible with the dramatic damage of the basal ganglia.

1. Dominique Laplane, Michel Baulac, et al., "Pure Psychic Akinesia with Bilateral Lesions of the Basal Ganglia," *Journal of Neurology, Neurosurgery and Psychiatry* (1984): 377–85.

The third is medical: the dysfunction of the different parts of the basal ganglia is at the origin not only of involuntary abnormal movements, but also of involuntary abnormal thoughts and emotions (i.e., touching upon the whole of psychiatry). Indeed, as discussed previously, a large number of psychiatric disorders are observed in patients with basal ganglia lesions,[4] including the rare and astonishing syndrome of "loss of psychic auto-activation (box 4.2). Because we assume that the basal ganglia contribute essentially to the mental function of subconsciousness, this strongly indicates that several mental disorders do result from the dysfunction of subconsciousness. Conversely, it also suggests that this essential mental function plays an important role in the existence of humans as individuals and as a community. Understanding the basal ganglia's function provides a new lens for viewing human behavior and addressing impairments, particularly in the field of psychiatry (box 4.3).

BOX 4.3: FROM THE GREAT SIGMUND TO MODERN PSYCHIATRY

What will be the future of psychiatry, which is still largely based on the doctrine of psychoanalysis? The psychoanalytic theory espoused by Sigmund Freud in the late nineteenth century influenced minds to the point of becoming a great movement with a major impact on the discipline of psychiatry. For a period in 1885–1886, Freud spent time in Jean-Martin Charcot's neurology service at Salpêtrière Hospital in Paris.

Charcot's research was based in his anatomo-clinical method of linking clinical signs with anatomical lesions. Initially, the patient's clinical symptoms were documented longitudinally; and, after death, an autopsy was performed to examine the brain and spinal cord. By combining clinical and anatomical data, Charcot and his followers were able to suggest concrete clinical-anatomical correlations. This method helped to define the pathways responsible for normal and abnormal neurological signs and was pivotal to a new classification of neurological diseases based on anatomy. After leaving Salpêtrière, Freud turned away from the anatomo-clinical method to theoretical neurology and psychology, a process that subsequently led to the birth of the now largely controverted theory of psychoanalysis. It is interesting to muse about the outcome had Freud had not taken this departure . . .

The difficulty in the field of psychiatry is that, in contrast to neurology, where loss of neurons and specific histological stigmata can be identified postmortem and to a lesser degree in vivo (neuroimaging), dysfunction of nerve cells is not observable in autopsy material, and even less in patients, as the brain tissue appears normal, although it does not function normally at the cellular level. The recent development of new technologies will likely elicit significant progress in the comprehension of the various mechanisms of most psychiatric disorders at the molecular and cellular levels.

5

Looking Deeper Into the Basement of the Brain

Is the Concept of Subconsciousness Presented Here Valid?

In the chaos of terminologies, a definition of subconsciousness and its subtypes is proposed, drawing in particular on clinical observation and the most recent data on brain physiology. Philosophers and psychologists who have studied subconsciousness, generally under the term "cognitive unconscious," have adopted definitions that go in a similar direction, but with significant nuances.[1] The advantage of my definition is to individualize subconsciousness in relation to other aspects of consciousness—literal metaconsciousness and consciousness—even though no clear distinction exists, because there is no discrete separation of these three faculties that indicates whether they fall on the automatic or non-automatic side. The reason is that we are constantly moving from one definition to another, just as the brain moves from one state of consciousness to another (figure 5.1).

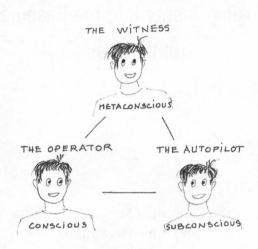

5.1 The states of consciousness.

As described earlier in the book, the concept of consciousness was divided into its multiple subtypes, intentional or not, and in the latter case with or without memory. It is logical to distinguish what is intentional and what is not within the concept of subconsciousness, even though this may appear to be an oversimplification. It seems self-evident that we keep the memory of an action when we intend to do so. However, this is not the case when there is no intention. The brain keeps a memory trace, either implicit and not recallable, as in infra-consciousness, or explicit, because it can be recalled, as in the case of the Freudian unconscious. Although the Freudian unconscious may seem to be underestimated, or even "desacralized," this is not the case. In the account presented here, Freud's unconscious is only reassigned without any concern for prioritization. This presentation of subconsciousness may seem artificial, as is any taxonomy, but it has the advantage of integrating the different forms of consciousness into a general framework, making it possible to seek the physiological foundations of each of its linked components. In this way, it allows the initiation of a scientific approach, one that identifies the sets of nerve cells that are specifically involved in each of the components of the subconscious.

Even if it is not a star in relation to its illustrious parents, metaconsciousness and consciousness, the mental autopilot that is subconsciousness has at least three merits:

1. *We are in a state of subconsciousness all our lives.* Subconsciousness alone has the ability to manage the individual's automatic adaptation to the environment (walking, pedaling, playing piano scales, etc.), provided there is no need to adapt to a new situation. However, subconsciousness is also present when we adapt to a new situation (consciousness) or when we engage in a task of reflection on our own thoughts (metaconsciousness).

2. *Subconsciousness is present in living beings that have an elaborate nervous system*, dominant in animals such as fish,

birds, and reptiles (probably absent in animals that have only a few neurons, such as earthworms). This says something about the power and robustness of this ability that has allowed the animal kingdom to survive to the present day.

3. Finally, because subconsciousness is a way of thinking, doing, feeling, and being that does not require our attention—we are truly unaware—this mental faculty expresses in real time the intimacy of our thoughts and emotions, that is, the deepest part of us. It is probably the best reflection of our personality. This shows the value of carefully examining our interlocutors when they are natural, without calculation or simulation, to discern who they really are, so that we may better understand and interact with them.

The term "subconscious" is only a word, certainly defined here as an overlearned automatic behavior, under the control of our nonautomatic consciousness and under the benevolent gaze of metaconsciousness. Is that really its only role? Couldn't highly sophisticated behaviors, such as making a decision, for example, be handled by this subconsciousness? This could be suggested by the experiences of Benjamin Libet,[2] who nearly forty years ago showed that a decision could be taken even before someone was aware of it (figure 5.2).

5.2 Subconsciousness and free will: the experience of Benjamin Libet. (a) You are placed in front of a clock with seconds scrolling in front of you. You have a button at your disposal that you can press at any time. You only need to remember the number indicated by the clock when you make your decision to press the button. During this experiment, electrodes placed on your skull to monitor your brain activity determine: (1) when you make the decision to press the button (decision); (2) when your cerebral cortex is activated when you make that decision (onset of brain activation); and (3) when you press the button (movement).

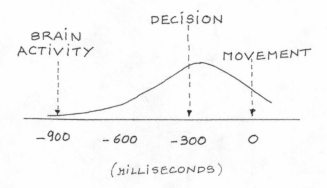

BRAIN
ACTIVITY

DECISION

MOVEMENT

−900   − 600   −300   0

(MILLISECONDS)

5.2 (b) As expected, the start of the movement takes place about 300 milliseconds after the conscious decision to press the button. Surprisingly, the onset of brain activity (measured by the electrical signal collected on the cerebral cortex) occurs 500 milliseconds before the decision to make the movement. Brain activation precedes conscious decision. In other words, you are aware of making a movement at some point, but your brain made the decision half a second earlier! If the results of this experiment are validated, it means that your brain has started to act, even though you have not yet decided. But if that is the case, you are not free to decide—because it is your brain that decides. Does that mean you do not have free will? Not really, because if consciousness is indeed the "operator," as suggested in this text, a decision (necessarily intentional) can be taken before becoming aware of it, that is, subconsciously. And if subconsciousness is really managed in a predominant way (like a "hub"), it would not be the cerebral cortex that would be predominantly responsible for it, but other areas of the brain that play a crucial role in our unconscious intentional behaviors. What are the cerebral structures that play a predominant role in subconsciousness, according to this book? Selective areas within the cerebral cortex? Any subcortical brain structures? Why not the basal ganglia, known to play a role in the genesis of subconsciousness? A hypothesis whose validity remains to be demonstrated.

Subconsciousness, the Societal Stakes

The point of presenting subconsciousness, with its subdivisions, including the Freudian unconscious, as the privileged interlocutor of consciousness and metaconsciousness, is to propose a general framework of thought and its behavioral expression in humans. Yet—it will be said—there has been no shortage of theories to explain the dynamics of human behavior.

Just over one hundred years ago, psychoanalytical thinking influenced minds to the point of becoming a great movement of philosophical reflection, with the sociological success that we know in Western countries, not to mention the major impact that it had on the discipline of psychiatry. It is true that the lack of scientific validation of psychoanalytical theories means that questions are raised as to whether that approach is not outdated. Imagine the reaction of the more conservative among us if they are faced with the proposition that the theory of Freud and his successors is now stored at the bottom of the subconscious closet! Without neglecting the phenomenal impact of psychoanalytical thinking in medicine, it is perhaps time to put the Freudian unconscious back in its rightful place: subconscious, nonintentional, not memorized, but eventually recallable.

Until about fifty years ago, many people considered the brain to be a type of black box that received information and produced action, yet whose internal functioning was considered too complex to worry about deciphering. The thought was, "After all, why would we need to know about the intimate functioning of the brain machine to interpret the origin of behaviors?" However, because it is precisely through their behaviors that we judge our children, friends, and fellow citizens, and because it has been well established that behaviors are produced by the brain, there was a growing belief that it would be difficult to understand human beings without knowing what was going on in the brain. It is true that you can drive your car without understanding how the engine works. However, the Formula 1 driver improves the performance of his or her car by participating in engine tuning with engineers. Similarly, physiologically decoding the function of the "brain" machine should provide valuable clues to the human and social sciences to better understand human behavior at the individual and community levels.

Hence the spectacular development of computational neuroscience coupled with neuroimaging to advance the understanding of information processing mechanisms within the brain. Today, the scientific developments summarized here show that understanding the anatomical and physiological bases of the human brain, and more particularly the basal ganglia, intimately connected to the cerebral cortex, allows us to better comprehend the foundations of the mental faculty that I propose to be the most important structure to explain human behavior—subconsciousness.

If the best reflection of subconscious brain function is expressed by what we perceive of human behavior in the basic state, without attention or awareness, this suggests that we must know how to trust semiology, that is, to observe and listen to individuals when they are natural, without mannerisms or being primed. The finest analysis of human behavior is possible when the mask has fallen—and all the more accessible for the psychologist and the physician.

Taking into account the concept of subconsciousness may also provide an opportunity to demystify what is believed to be the unequaled power of human mental functions. They are certainly sophisticated, but perhaps not as sophisticated as we think, if we compare ourselves with animals. Although nonhuman animals lack the amazing abilities of *Homo sapiens*, who dominate the animal world through their spoken language, manual dexterity, elaborate strategic and anticipatory capabilities, and their ability to process and transmit abstract concepts, they do have intellectual and emotional faculties that we often pretend to ignore by using anthropomorphism and anthropocentrism. For example, their unsuspected capacities of communication without language, the sense of family and community, love and fidelity, the reactivity of the senses and the cunning to escape from a predator, the ability to live as a couple or in society. In short, all these capacities of adaptation to the environment that have ensured their survival over thousands of millennia. And yet, these qualities in animals remain, for the most part, in the realm of subconsciousness. What about humans (figure 5.3)?

Human behavior is encumbered by habits and routines that are gradually internalized within the family, at school, in the community. This happens automatically, without our

5.3 Animals, man, and the subconscious.

thinking or becoming aware of it. They are natural ways of being, perceiving, and acting ("Hello, how are you?" . . . "Would you like some more?" . . . "Please pass me the salt." . . . "Thank you." . . . "It looks like it's going to rain." . . . "Be careful, the stove is hot." . . . " Our Father who is in heaven . . ." . . . "Ladies and gentlemen . . . It is with a heavy heart that I pay tribute to our dear . . .") which constitute part of the concept of subconsciousness.

Taking into account this concept of subconsciousness is therefore essential to interpret and predict human behavior at the individual or societal level. The experimental observations reported here on the subject of subconsciousness are therefore entirely justified in enriching sociological reflection. The challenge of studying subconsciousness at all levels, from biology to behavior, is therefore also sociological. What would be the use of robust but dry cognitive research that is not applicable? What would be the use of sparkling but limp psychological approaches that have no scientific basis? The faculty of subconsciousness is obviously vital for most animals that have only subconsciousness, but also for humans, who need these routines to survive, unless we are hermits, depressed, or mad. At the same time, subconsciousness as such should not be valued in isolation. Indeed, it is a key revealer of the nature of people and an essential support of human personality. What would we be, if we were nothing more than subconsciousness? Evolved beasts, without finesse, without artistic emotion, without high-reaching thought, without the capacity for abstraction or anticipation, without creativity, without ethics—in other words, without meaning. But it is the subconscious that makes the other attributes possible. To overlook it is to ignore the foundation of who we are and potentially invaluable avenues for healing.

Notes

1. The Interrelated Levels of Consciousness

1. Marcel Gauchet, *L'inconscient cérébral* (Paris: Le Seuil, 1992).
2. Daniel Dennett, *Brainstorms: Philosophical Essays on Mind and Psychology* (Cambridge, Mass.: MIT Press, 1978).
3. Gordon Gallup, "Self-Recognition in Primates," *American Psychology* 32 (1977): 329–38.
4. Antonio Damasio, *The Error of Descartes* (Paris: Odile Jacob, 1999).
5. William James, *Précis de psychologie: Les empêcheurs de tourner en rond* (Paris: Le Seuil, 2002).
6. Pierre Buser, *L'inconscient aux mille visages* (Paris: Odile Jacob, 2005).
7. *Petit Robert Dictionnaire*, nouvelle édition (Paris: Le Robert, 2008).
8. Buser, *L'inconscient aux mille visages*.
9. Pierre Janet, *Evolution de la personnalité* (Paris: Chahine, 1929).

2. The Complexities of Subconsciousness

1. Yves Agid, *L'homme subconscient* (Paris: Robert Laffont, 2013).
2. Jean de la Fontaine, "Les deux pigeons," Book 9, fable 2 (Paris: Charpentier, 1709).
3. Lawrence Weiskrantz, *Blindsight: A Case Study and Implications* (Oxford: Oxford University Press, 1988).
4. Stanislas Dehaene et al., "Cerebral Mechanisms of Word Masking and Unconscious Repetition Priming," *Nature Neuroscience* 4, no. 7 (2001): 752–58.
5. Jean-Martin Charcot, *Oeuvres complete: Leçons sur les maladies du système nerveux recueillies et publiées par Bourneville* (Paris: L. Bataille, 1892).
6. Max Wertheimer, *Productive Thinking* (New York: Harper and Brothers, 1959).
7. Henri Poincaré, *L'invention mathématique* (Paris: Jacques Gabay, 1993).

3. A Physiological Base for the Subconscious in the Brain

1. David Marsden, "The Mysterious Motor Function of the Basal Ganglia," *Neurology* (1982): 514–39.
2. Jonathan Mink, "The Basal Ganglia: Focused Selection and Inhibition of Competing Motor Programs," *Progress Neurobiology* 50 (1996): 381–425.
3. Ann Graybiel, "The Basal Ganglia: Learning New Tricks and Loving It," *Current Opinion Neurobiology* 15 (2005): 638–44; Stephane Lehéricy et al., "Distinct Basal Ganglia Territories Are Engaged in Early and Advanced Motor Sequence Learning," *Proceedings of the National Academy of Sciences USA* 102 (2005): 12566–71.

4. Kalash Bathia and David Marsden, "The Behavioral and Motor Consequences of Focal Lesions of the Basal Ganglia in Man," *Brain* 117 (1994): 859–76, doi: 10.1093/brain/117.4.859.

5. David Grabli et al., "Behavioral Disorders Induced by External Globus Pallidus Dysfunction in Primates: I. Behavioural Study," *Brain* 127 (2004): 2039–54, doi: 10.1093/brain/awh220.

6. Boulos Bejjani et al., "Transient Acute Depression Induced by High Frequency Deep Brain Stimulation," *New England Journal of Medicine* 340 (1999): 1476–80, doi: 10.1056/NEJM 199905133401905.

7. Luc Mallet et al., "Stimulation of Sub-territories of the Sub-thalamic Nucleus Reveals Its Role in the Integration of the Emotional and Motor Aspects of Behavior," *Proceedings of the National Academy of Sciences USA* 104, no. 25 (2007): 10661–66, doi: 10.1073/pnas.0610849104.

4. The Diseased Subconscious and Implications for the Future

1. Patricia Limousin, Pierre Pollak, et al., "Effect on Parkinsonian Signs and Symptoms of Bilateral Subthalamic Nucleus Stimulation," *Lancet* (1995): 90–95.

2. Marie Vidailhet, Laurent Vercueil, et al., "Bilateral Deep Brain Stimulation of the Globus Pallidus in Primary Generalized Dystonia," *New England Journal of Medicine* (2005): 459–67.

3. Marie-Laure Welter, Luc Mallet, et al., "Internal Pallidal and Thalamic Stimulation in Patients with Tourette Syndrome," *Archives of Neurology* (2008): 952–57.

4. Kalash Bathia and David Marsden, "The Behavioral and Motor Consequences of Focal Lesions of the Basal Ganglia in Man," *Brain* 117 (1994): 859–76.

5. Looking Deeper Into the Basement of the Brain

1. Stanislas Dehaene, *Le code de la conscience* (Paris: Odile Jacob, 2014); Daniel Dennett, *Brainstorms: Philosophical Essays on Mind and Psychology* (Cambridge, Mass.: MIT Press, 1978); Pierre Janet, *Evolution de la personnalité* (Paris: Chahine, 1929).

2. Benjamin Libet et al., "Time of Conscious Intention to Act in Relation to Onset of Cerebral Activity (Readiness Potential): The Unconscious Initiation of a Freely Voluntary Act," *Brain* 106, no. 3 (1983): 623–42, doi: 10.1093/brain/106.3.623.

Further Reading

Bartels, Andreas, and Semir Zeki. "The Neural Basis of Romantic Love." *Neuroreport* (2000): 13829–34. doi: 10.1097/00001756-200011270-00046.

Bergson, H. *Matière et mémoire: Essai sur la relation du corps et de l'esprit.* Paris: Presses Universitaires de France, 1965.

Buser, Pierre. *L'inconscient aux mille visages.* Paris: Odile Jacob, 2005.

Carter, Rita. *The Human Brain Book.* New York: Dorling Kindersley, 2009.

Chalmers, D. *The Conscious Mind.* New York: Oxford University Press, 1996.

Churchland, P. M. *A Neurocomputational Perspective: The Nature of Mind and the Structure of Science.* Cambridge, Mass.: MIT Press, 1990.

Damasio, Antonio. *The Error of Descartes.* Paris: Odile Jacob, 1999.

de La Mettrie, J. O. *L'homme machine.* Paris: J. J. Pauvert, 1966.

Dennett, Daniel. *Elbow Room: The Variety of Free Will Worth Wanting.* Cambridge, Mass.: MIT Press, 1984.

Edelman, G. *Neural Darwinism.* New York: Basic Books, 1987.

Fourneret, P., and M. Jeannerod. "Limited Conscious Monitoring of Motor Performance in Normal Subjects." *Neuropsychologia* 36 (1998): 1133–40.

Freud, Sigmund. *Der Witz und seine Beziehung zum Unbewusten*. Paris: Gallimard, 1988.

Gauchet, Marcel. *L'inconscient cérébral*. Paris: Le Seuil, 1992.

Graybiel, Ann. "The Basal Ganglia: Learning New Tricks and Loving It." *Current Opinion Neurobiology* 15 (2005): 638–44.

James, W. "Does Consciousness Exist?" *Journal of Philosophy, Psychology, and Scientific Methods* (1904): 477–91.

Kahneman, D. *Thinking, Fast and Slow*. New York: Farrar, Straus and Giroux, 2011.

Laplane, Dominique, Michel Baulac, et al. "Pure Psychic Akinesia with Bilateral Lesions of the Basal Ganglia." *Journal of Neurology, Neurosurgery, and Psychiatry* (1984): 377–85.

Lehéricy, Stephane, et al. "Distinct Basal Ganglia Territories Are Engaged in Early and Advanced Motor Sequence Learning." *Proceedings of the National Academy of Sciences USA* 102 (2005):12566–71.

Lhermitte, François, and Michel Serdaru. "Unconscious Processing Memory Recall: A Study of Three Amnestic Patients." *Cortex* 30 (1993): 35–42. doi: 10.1016/s0010-9452(13)80209-2.

Maine de Biran. *Influence de l'habitude sur la faculté de penser*. Book 3. Tisserand, 1803.

Mallet, Luc, Mircea Polosan, et al. "Subthalamic Nucleus Stimulation in Severe Obsessive-Compulsive Disorder." *New England Journal of Medicine* (2008): 2121–34.

Posner, M. I. "Attention: The Mechanisms of Consciousness." *Proceedings of the National Academy of Sciences USA* 91: 7398–7403.

Ribot, T. *La psychologie anglaise contemporaine*. Paris: L'Harmattan, 1887.

Index

Page numbers in *italics* represent figures or tables.